KV-038-672
ANDERSONIAN
WITHDRAWN
FROM
LIBRARY
STOCK
UNIVERSITY OF STRATHCLYDE

TREATMENT OF
HAZARDOUS WASTE LEACHATE

TREATMENT OF HAZARDOUS WASTE LEACHATE

Unit Operations and Costs

by

J.L. McArdle, M.M. Arozarena, W.E. Gallagher

PEI Associates, Inc.
Cincinnati, Ohio

NOYES DATA CORPORATION

Park Ridge, New Jersey, U.S.A.

Library of Congress Catalog Card Number 87-34715
ISBN: 0-8155-1160-4
ISSN: 0090-516X
Printed in the United States

Published in the United States of America by
Noyes Data Corporation
Mill Road, Park Ridge, New Jersey 07656

10 9 8 7 6 5 4 3 2 1

Library of Congress Cataloging-in-Publication Data

McArdle, J.L.
Treatment of hazardous waste leachate.

(Pollution technology review, ISSN 0090-516X ;
no. 151)
Bibliography: p.
Includes index.
1. Hazardous wastes. 2. Hazardous waste sites--
Leaching. 3. Leachate. I. Arozarena, M.M. (Michael M.)
II. Gallagher, W.E. (William E.) III. Title.
IV. Series.
TD811.5.M39 1988 628.4 87-34715
ISBN 0-8155-1160-4

Foreword

This book describes twenty technologies that have been employed to treat leachates from hazardous waste sites. It will be useful to industrial and engineering firms which must deal with leachate treatment.

The U.S. Environmental Protection Agency's hazardous waste site cleanup program, referred to as Superfund, was authorized and established in 1980 by the enactment of the Comprehensive Environmental Response, Compensation, and Liability Act (CERCLA). This legislation allows the federal government (and cooperating state governments) to respond directly to releases and the threat of releases of hazardous substances and pollutants or contaminants that could endanger public health or welfare or the environment. Leachate is one type of release covered by this law. It is formed when water percolates through a waste-disposal site, and, if not properly contained and collected, it can threaten the local hydrogeologic environment. The objective of this book is to provide guidance in the treatment of hazardous waste leachate.

The twenty unit operations are reviewed for their applicability to the treatment of hazardous waste leachate. They are classified into the following four categories: pretreatment operations, physical/chemical treatment processes, biological treatment, and post-treatment operations. Typical treatment process trains (combinations of unit operations) are presented for leachate containing organic and/or inorganic contaminants.

Factors affecting leachate generation and its compositions are addressed. Also discussed is the management of residuals—sludges, air emissions, concentrated liquid waste streams, spent carbon—generated by the various treatment techniques.

The information in the book is from *A Handbook on Treatment of Hazardous Waste Leachate,* prepared by J.L. McArdle, M.M. Arozarena, and W.E. Gallagher of PEI associates, Inc. for the U.S. Environmental Protection Agency, Dec. 1986.

The table of contents is organized in such a way as to serve as a subject index and provides easy access to the information contained in the book.

Advanced composition and production methods developed by Noyes Data Corporation are employed to bring this durably bound book to you in a minimum of time. Special techniques are used to close the gap between "manuscript" and "completed book." In order to keep the price of the book to a reasonable level, it has been partially reproduced by photo-offset directly from the original report and the cost saving passed on to the reader. Due to this method of publishing, certain portions of the book may be less legible than desired.

ACKNOWLEDGMENTS

This document was prepared for EPA's Office of Research and Development, Hazardous Waste Engineering Research Laboratory, by PEI Associates, Inc., under subcontract to Battelle Columbus Laboratories, Mr. Edward J. Opatken served as the EPA Technical Project Monitor. Mr. Ron Clark of Battelle provided contract management assistance and project overview. PEI's efforts were directed by Mr. Jack S. Greber and managed by Ms. Judy L. McArdle, Messrs. Thomas C. Ponder and Michael C. Hessling provided senior technical review. Other major contributors include Messrs. Michael M. Arozarena, William E. Gallagher, Jack P. Paul, John E. Spessard, and Ms. Martha H. Phillips.

The authors wish to acknowledge the valuable contribution of the many individuals who conducted tours of their leachate treatment facilities or provided other technical input. These individuals include Messrs. Nicholas P. Kolak and Brian Sadowski (New York State Department of Environmental Conservation); Mr. Brian Ullensvang (U.S. Environmental Protection Agency, Region IX); Messrs. Samuel M. Lybrand and George Widmann (G.R.O.W.S., Inc.); Messrs. Philip A. Herzbrun and Jack Curtis (CECOS International, Inc.); Mr. Ron Peterson (Bofors-Nobel, Inc.); Ms. Sandy K. Henderson and Mr. David Van Wyk (Dow Chemical Texas Operations); Mr. Allan R. Wilhelmi (Zimpro, Inc.); and Mr. Fred Mendicino (Calgon Carbon Corp.).

NOTICE

Contents and Subject Index

1. Introduction

1.1 BACKGROUND AND OBJECTIVES

The U.S. Environmental Protection Agency (EPA) hazardous waste site cleanup program, referred to as Superfund, was authorized and established in 1980 by the enactment of the Comprehensive Environmental Response, Compensation, and Liability Act (CERCLA), Public Law (PL) 96-510. This legislation allows the Federal government (and cooperating State governments) to respond directly to releases and the threat of releases of hazardous substances and pollutants or contaminants that could endanger public health or welfare or the environment. Prior to the passage of PL 96-510, Federal authority with regard to hazardous substances was mostly regulatory in nature through the Resource Conservation and Recovery Act (RCRA) and the Clean Water Act and its predecessors.

Public Law 96-510 and the regulations based on it not only govern accidental releases that may occur from time to time, but also releases that already have taken place and continue to take place from uncontrolled waste-disposal sites. Leachate is one type of release covered by this law. It is formed when water percolates through a waste-disposal site, and if not properly contained and collected, it can threaten the local hydrogeologic environment. The objective of this handbook is to provide guidance in the treatment of hazardous waste leachate.

1.2 ORGANIZATION OF THE HANDBOOK

Subsequent sections of this document are concerned with the nature of hazardous waste leachate and applicable methods of treatment. Section 2 addresses the factors that affect leachate generation, and Section 3 addresses its composition. The treatability of leachate constituents is covered in Section 4, which presents a process applicability matrix that can be used to determine the appropriateness of a given technology for altering or removing hazardous leachate constituents and to determine pretreatment requirements. Section 5 describes the selection and combining of unit operations to form an efficient treatment process train. Examples of process trains for removing inorganic and organic constituents from leachate at Superfund sites are presented. Section 6 gives details concerning the applicability/limitations, design considerations, performance, and capital and annual operating and maintenance costs of 20 unit processes with demonstrated or potential applicability to the treatment of hazardous waste leachate. The technologies are classified as pretreatment operations, physical/chemical treatment operations, biological treatment operations, and post-treatment operations. The order in which the technologies within each category are presented reflects the reliability of the process for leachate treatment applications (i.e., technologies that have been widely demonstrated are presented first; innovative technologies or technologies that have not been demonstrated with hazardous waste leachate are presented last). Section 7 discusses management of residuals generated by the various leachate treatment technologies, including chemical/biological sludges, air emissions of volatile organic compounds, concentrated liquid waste streams, and spent carbon.

2. Leachate Generation

2.1 MECHANISM OF LEACHATE GENERATION

Leachate is generated by the movement of water through a waste-disposal site. Figure 1 illustrates the mechanism of leachate generation, which is tied to the hydrologic cycle. Precipitation falling on the land surface will either infiltrate the cover soil or leave the site as surface runoff, depending on surface conditions. Infiltrated water that is not subsequently lost by evapotranspiration or retained as soil moisture will percolate down through the waste deposit. Initially, this liquid will be absorbed by the waste material. When the field capacity (moisture-retention capacity) of the waste or portions of the waste is exceeded (which may take from several months to several years), leachate will be produced. At waste-disposal sites with no provisions for collection, this leachate can contaminate underlying ground-water aquifers or nearby surface streams.

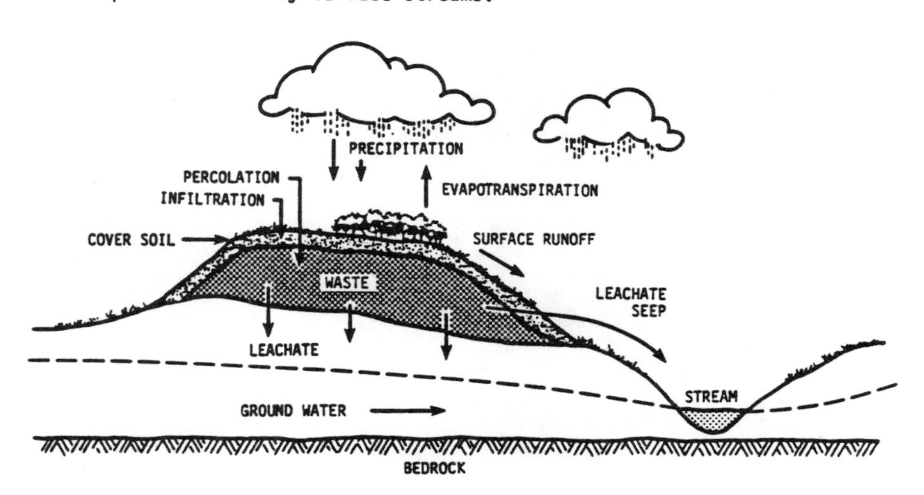

Figure 1. Mechanism of leachate generation.

2.2 FACTORS AFFECTING LEACHATE GENERATION

Leachate generation (flow) varies greatly from site to site and over time at the same site. Among the many factors that contribute to this variability are the local climate and meteorology, site topography, cover soil and vegetation, and site hydrogeology (Mikucki et al. 1981).

Climate and meteorology (rainfall, temperature, humidity, windspeed) determine the availability of water for leachate production. Leachate production potential is highest in areas where the precipitation is high and the actual evapotranspiration is low. Seasonal variations in leachate production correspond to local patterns of precipitation and infiltration (freeze/thaw). Peak flows are typically associated with spring thaws.

Site topography affects surface runoff patterns and the quantity of water available for infiltration. Flat surfaces have the highest capacity for infiltration.

Net infiltration is a function of both the type of cover soil (particularly its permeability and moisture retention capacity) and the kind and amount of surface vegetation (which determine the amount of soil water lost because of evapotranspiration). Unauthorized or abandoned waste-disposal sites will produce greater quantities of leachate than properly designed and permitted landfills that have been capped and seeded.

Site hydrogeologic characteristics such as depth to water table and the ground-water flow regime influence the extent of ground-water intrusion into the disposal site. Although ground-water intrusion increases the volume of leachate generated, it also dilutes the strength of the leachate stream.

2.3 DESIGN IMPLICATIONS

On the average, leachate is generated in low to moderate flows (<100,000 gal/day); however, seasonal and day-to-day fluctuations in leachate volume can have a significant impact on the design of a leachate treatment plant. With continuous treatment operations, some form of flow equalization is normally required to handle peak flows and optimize plant performance. Processes that can be operated intermittently have the advantage of being able to meet increases or decreases in treatment demands over the life of the plant.

3. Leachate Characteristics

3.1 COMPOSITION OF LEACHATE

As water percolates through a waste deposit, it solubilizes (leaches) various components of the waste and becomes polluted. This leachate typically exhibits high concentrations of dissolved organics (BOD_5, COD, TOC), toxics (TOX), and metals; high color, odor, and turbidity; and low pH. Table 1 summarizes analytical data from 30 uncontrolled hazardous waste sites on the composition of ground and surface waters contaminated by leachate. Although these data do not adequately reflect the strength of raw leachate (wherein no dilution has occurred), they do illustrate the widely varying composition of leachate from site to site. This variability is also typical of data collected from one site over a long period of time (Mikucki et al. 1981).

3.2 FACTORS AFFECTING LEACHATE COMPOSITION

The factors that have the greatest effect on leachate composition are those that influence the degradation of the waste and those that affect the mobilization of waste components and degradation products.

Buried wastes are degraded both chemically and biologically. Chemical oxidation of organic and inorganic components is largely a function of the oxidation-reduction potential of the waste. Biological decomposition of organic matter is dependent on a viable population of microorganisms, which in turn is dependent on the composition of the waste, concentration of toxic constituents, availability of nutrients, oxygen levels, temperature, moisture, and pH. Initially, biological decomposition proceeds aerobically. As oxygen is depleted, however, aerobic microorganisms give way to anaerobes. Because anaerobic decomposition occurs very slowly, degradation of organic contaminants may require several decades. Although leachate production will continue for several years, the strength of the leachate stream will gradually decrease with time (Chian and DeWalle 1977).

Mobilization of leachate constituents is a function of chemical solubility and the rate of water movement through the waste. Solubility is controlled by the solution pH; under acidic conditions, the solubility of many metals increases. The rate of water movement through the disposal site determines the length of time the water is in contact with the waste and the ultimate strength of the leachate.

TABLE 1. CONVENTIONAL AND PRIORITY POLLUTANT ANALYSES OF CONTAMINATED GROUND AND SURFACE WATERS IN THE VICINITY OF UNCONTROLLED HAZARDOUS WASTE SITES[a]

Conventional pollutants	Range, mg/liter	No. of sites
Biochemical oxygen demand (5-day) (BOD_5)	42-10,900	3
Chemical oxygen demand (COD)	24.6-41,400	6
Total organic carbon (TOC)	10.9-8,700	8
Oil and grease	90	1
Total suspended solids (TSS)	<3-1,040	4
Total dissolved solids (TDS)	1,455-15,700	4
Specific conductance, µmhos/cm	80-2,000	2
pH, std. units	3-7.9	7
Alkalinity (as $CaCO_3$)	20.6-5,400	3
Hardness (as $CaCO_3$)	700-4,650	2
Total Kjeldahl nitrogen	<1-984	4
Ammonia-nitrogen	<0.010-1000	3
Nitrate-nitrogen	0.01-0.10	3
Nitrite-nitrogen	<0.01-0.10	2
Total phosphorus	<0.1-3.2	2
Sulfate	1.2-505	4
Sulfide	<0.1	1
Chloride	3.65-9,920	6
Calcium	164-2,500	4
Magnesium	25-453	3
Sodium	4.6-1,350	5
Potassium	6.83-961	3

(continued)

TABLE 1 (continued)

Priority pollutants	Range, µg/liter	No. of sites
Volatile compounds		
Benzene	<1.1-7,370	5
Bromoform	0.2	1
Carbon tetrachloride	<1-25,000	3
Chlorobenzene	4.6-4,620	5
Chlorodibromomethane	3.9	1
Chloroform	0.02-4,550	4
Dichlorobromomethane	ND[b]-35	1
1,1-Dichloroethane	<5-14,280	2
1,2-Dichloroethane	2.1-4,500	5
1,1-Dichloroethylene	28-19,850	5
1,2-Dichloropropane	<22	1
1,3-Dichloropropylene	c	1
Ethylbenzene	3-10,115	4
Methylene chloride	300-184,000	3
1,1,2,2-Tetrachloroethane	<5-1,590	1
Tetrachloroethylene	ND-8,200	5
Toluene	<5-100,000	7
1,1,1-Trichloroethane	1.6-590,000	5
1,1,2-Trichloroethane	<5-870	2
Trichloroethylene	<3-260,000	4
Trichlorofluoromethane	<5-18	1
Vinyl chloride	140-32,500	1
Acid-extractable compounds		
2-Chlorophenol	<3-48	1
2,4-Dinitrophenol	10-99	2
2-Nitrophenol	8,600-12,000	1
Pentachlorophenol	2,400	1
Phenol	<3-17,000	4

(continued)

TABLE 1 (continued)

Priority pollutants	Range, µg/liter	No. of sites
Base/neutral-extractable compounds		
Anthracene	<10-670	1
Bis(2-ethylhexyl) phthalate	53,000	1
Di-n-butyl phthalate	21,732	1
Hexachlorobenzene	32-<100	1
Hexachlorobutadiene	<20-109	2
Hexachlorocyclopentadiene	<100	1
Naphthalene	<10,000-18,698	2
Nitrobenzene	ND-740	1
1,2,4-Trichlorobenzene	<10-28	2
Pesticides/PCB's		
Aldrin	<10	2
Dieldrin	<2-4.5	1
Endrin	<2-9	1
Heptachlor	573	1
Archlor 1245	70	1
Inorganics		
Antimony	2000	1
Arsenic	11-<10,000,000	6
Beryllium	7	1
Cadmium	5-8,200	6
Chromium	1-208,000	7
Copper	1-16,000	9
Cyanide	0.5-14,000	2
Lead	1-19,000	6
Mercury	0.5-7.0	7
Nickel	20-48,000	4
Selenium	3-590	4
Silver	1-10	2

[a] Source: Shuckrow, Pajak, and Touhill 1982.

[b] ND = not detected.

[c] Present, but not quantified.

3.3 DESIGN IMPLICATIONS

The chemical and physical characteristics of leachate are the primary considerations in the design of a treatment system. The technologies applicable to hazardous waste leachate treatment are essentially the same as those applied to municipal wastewater and contaminated ground-water treatment; however, hazardous waste leachate is typically more concentrated than municipal wastewater or ground water, and multistage treatment is often required. High-strength leachates are frequently treated by low-cost processes capable of removing gross contamination and then followed by more costly processes to remove the residuals or to "polish" the effluent. The most common application of this design strategy involves the use of air stripping to remove greater than 90 percent of the volatile organics, followed by granular activated carbon adsorption to remove the residual organics.

Typically, hazardous waste leachate is composed of multiple chemical contaminants, including both organic and inorganic compounds. As a result, treatment systems that involve several different chemical, physical, and biological operations are normally required. The proper combination of these processes is one of the most important considerations in the design of a treatment system. For example, biological processes frequently are preceded by metals precipitation for protection of the microorganisms.

Some hazardous waste leachates require extensive pretreatment before they can be treated by primary physical, chemical, or biological methods. Immiscible liquids, high turbidity, and corrosivity are all aspects of some leachates that must be handled during pretreatment to assure the effectiveness of the primary treatment operations. Extensive sampling and characterization of leachate to identify constituents, concentrations, and fluctuations are critical to the successful design of any system for treating hazardous waste leachate.

4. Treatability of Leachate Constituents

4.1 LEACHATE CHARACTERIZATION

Leachate characterization studies are designed to ascertain the type and concentration of constituents in the waste stream as well as the magnitude of variations in leachate flow rate and strength. Data from leachate characterization studies are useful in the screening of potentially applicable treatment technologies and as a baseline for evaluating the effectiveness of selected technologies. The following parameters are typically characterized (Shuckrow, Pajak, and Touhill 1982):

- Temperature
- Electrical conductivity
- Turbidity
- Settleable solids
- Suspended solids
- Total dissolved solids
- Volatile solids
- Oils, greases, and immiscible liquids
- Odor
- pH
- Oxidation-reduction potential (ORP)
- Acidity
- Alkalinity
- Biochemical oxygen demand (BOD)
- Chemical oxygen demand (COD)
- Total organic carbon (TOC)
- Specific organic compounds
- Heavy metals
- Other specific inorganic compounds
- Nitrogen and phosphorus compounds
- Dissolved oxygen
- Volatile organic acids
- Flow
- Toxicity

Characterization studies should be performed with leachate composited from several areas of the site. Possible sampling points include surface leachate seeps and ground-water monitoring wells installed in and around the waste deposit.

4.2 PROCESS APPLICABILITY SCREENING

When the characteristics of a particular leachate stream have been ascertained, potentially applicable processes for conversion or removal of target contaminants can be identified from the matrix in Figure 2. Each block of the matrix contains a "+", an "o", or a "-". Reading down a column for a contaminant of interest indicates which processes are effective in removing that contaminant (+). Reading across the row for a technology indicates the constituents that must be removed by pretreatment (-) to assure satisfactory

Technology	Suspended solids	Oil, grease, immiscible liquids	pH (acidic, basic)	Total dissolved solids	Metals	Cyanides	Volatile organics	Semivolatile organics	Pesticides, PCB's	Pathogens
Sedimentation	+	+	o	o	o	o	o	o	o	o
Granular-media filtration	+	−	o	o	o	o	o	o	o	o
Oil/water separation	o	+	o	o	o	o	o	o	o	o
Neutralization	o	o	+	o	o	o	o	o	o	o
Precipitation/flocculation/ sedimentation	+	+	o	+	+	o	o	o	o	o
Oxidation/reduction	−	−	o	o	+	+	o	+	+	+
Carbon adsorption	−	−	o	o	+	+	+	+	+	+
Air stripping	−	−	o	o	o	o	+	o	o	o
Steam stripping	−	−	−	o	o	o	+	o	o	o
Reverse osmosis	−	−	−	+	+	+	o	o	+	+
Ultrafiltration	−	−	−	+	+	o	o	+	+	+
Ion exchange	−	−	o	+	+	+	o	o	−	o
Wet-air oxidation	o	o	o	o	+	+	+	+	o	o
Activated sludge	−	−	−	o	−	o	+	+	o	o
Sequencing batch reactor	−	−	−	o	−	o	+	+	o	o
Powdered activated carbon treatment (PACT)	+	−	−	o	−	o	+	+	+	o
Rotating biological contactor	o	−	−	o	−	o	+	+	o	o
Trickling filter	−	−	−	o	−	o	+	+	o	o
Chlorination	o	−	−	o	o	+	o	o	o	+

Key: (+) process is applicable for removal of the contaminant; (o) process is not applicable for removal of the contaminant; (-) process is not applicable unless the leachate is pretreated for removal of the contaminant.

Figure 2. Process applicability matrix.

performance of that technology. For example, volatile organics can be removed from leachate by air stripping; however, suspended solids and oil and grease (which cause plugging of the packed bed) should be removed by pretreatment. Constituents that are neither removed by the technology nor require removal by pretreatment prior to application of the technology are indicated by an "o".

The process applicability matrix can be used to screen potential treatment technologies for their applicability to leachates whose compositions are known. Both process applicability and limitations are addressed in depth in the technology profiles in Section 6. The reader should not try to apply the matrix without consulting these discussions.

4.3 TREATABILITY STUDIES

Treatability studies provide guidance in the selection of the most cost-effective treatment alternative from among the potentially applicable technologies for a combination of leachate constituents. These studies examine the actual effectiveness of alternative methods as well as define design and operating standards. Because most leachate treatment technologies are adaptations of methods commonly used to treat wastewater and contaminated ground water, treatability studies can also identify process modifications that may be required because of the high strength and variable nature of hazardous waste leachate.

Treatability studies can be divided into two groups--bench-scale and pilot-scale--which differ in purpose, scale, cost, time, and leachate volume required, as summarized in Table 2. Although the distinction is not always clear, bench-scale studies are generally used for the preliminary evaluation and selection or rejection of the most promising treatment technologies, whereas pilot-scale studies are generally used to develop and optimize design and operating parameters of the selected process(es). As in the case of characterization studies, treatability studies should be performed with leachate composited from several areas of the site.

4.3.1 Bench-Scale Studies

Bench-scale studies are used to determine the impact of process variables on treatment performance over the range of expected operating conditions. The small scale and low cost of bench tests enable the evaluation of many variables. Examples of parameters that can be determined through bench-scale testing include chemical dosages, reaction rates, and optimum temperature, pressure, and pH. Certain parameters, however, cannot be tested at the bench-scale level, e.g., the effect of the size and configuration of the equipment on energy and mass transfer rates and the long-term stability of the process (EPA 1985a). Depending on the data requirements of the study (i.e., whether the data are to be used to develop pilot-scale tests or to design a full-scale unit) and the number of variables to be tested, bench-scale studies may require from a few days to a few months to complete.

TABLE 2. COMPARISON OF BENCH- AND PILOT-SCALE TREATABILITY STUDIES[a]

Parameter	Bench-scale studies	Pilot-scale studies
Purpose	Optimize process variables	Define design and operating criteria
Size	Laboratory or bench top	1 to 100 percent of full-scale
Quantity of leachate required	Limited amounts	Large amounts
Number of variables that can be considered	Many	Few
Time required	Days to months	Months to years
Typical cost range	0.5 to 2 percent of capital costs	2 to 5 percent of capital costs
Most frequent location	Laboratory	On site
Limiting considerations	Wall and boundary effects; volume effects; solids processing difficult to simulate	Limited number of variables; waste volume required; safety, health, and other risks

[a] Source: Adapted from EPA 1985a.

4.3.2 Pilot-Scale Studies

Pilot-scale studies are used to define design and operating criteria for a selected alternative and to demonstrate the long-term stability of the process. Pilot studies are typically performed under operating conditions approximating actual field conditions and in a scaled-down version of the permanent installation. Parameters that are more easily tested at pilot scale include mixing, separation, gas transfer, corrosion, and weather effects (EPA 1985a). The stability of the process under fluctuating influent conditions (flow rate and composition) should also be investigated to ensure that effluent limitations can be met. Pilot-scale studies may require from several months to a year or more to complete; this includes the time required for biological systems to become acclimated and for adsorption systems to break through.

5. Leachate Treatment Process Train Selection

Treatment of hazardous waste leachate is complicated by the diversity of organic and inorganic constituents that it contains. To effect a high degree of treatment efficiency requires several unit operations with specific applications and limitations (described subsequently in Section 6). This section describes waste- and site-specific factors that influence selection and integration of these processes. It also presents examples of process trains used to treat leachate containing inorganic contaminants, organic contaminants, and a combination of both.

5.1 SELECTION CRITERIA

Because the characteristics of hazardous waste leachate vary considerably from one site to the next, selection and integration of unit treatment processes are highly site-specific. Among the factors that influence selection are:

- Effluent discharge alternatives/limitations
- Treatment process residuals
- Permit requirements
- Cost-effectiveness of treatment

Treated leachate may be discharged to surface water, ground water (flushing), or a publicly owned treatment works (POTW). Surface-water or ground-water discharge requires complete treatment of the leachate stream, whereas discharge to a POTW entails only pretreatment. In general, effluent limitations will be less stringent in the latter case, although specific discharge limitations will vary from site to site.

Management of process residuals generated by the various leachate treatment alternatives can significantly add to the cost of a technology. Contaminants originally present in the leachate are concentrated in the residuals, which require further treatment and/or disposal. Methods for handling chemical/biological sludges, air emissions of volatile organic compounds (VOC's), concentrated liquid waste streams, and spent carbon are addressed fully in Section 7.

Certain aspects of all leachate treatment operations will be regulated under environmental laws such as the Clean Water Act, the Clean Air Act, and the Resource Conservation and Recovery Act. The time for and expense of obtaining all applicable operating and discharge permits should be factored into the treatment train selection process.

Ultimately, the treatment process(es) of choice at Superfund sites will be selected in accordance with Section 300.68(j) of the National Contingency Plan, which specifies that the appropriate extent of remedy is the "lowest cost alternative that is technologically feasible and reliable and that effectively mitigates and minimizes damage to and provides adequate protection of public health, welfare, and the environment." Table 3 compares estimated capital and annual operating and maintenance (O&M) costs of the treatment technologies covered in Section 6. These costs include the cost of residuals management.

5.2 LEACHATE CONTAINING INORGANIC CONTAMINANTS

Leachate containing primarily inorganic contaminants can be treated by a combination of physical/chemical processes. A typical process train might include equalization, oxidation/reduction, precipitation/flocculation/sedimentation, neutralization, and granular-media filtration. This process train is effective for removing most metals, including hexavalent chromium and soluble metal-cyanide complexes. Hexavalent chromium, which resists precipitation, can first be reduced to the trivalent state by the addition of sulfur dioxide or sodium bisulfite under acidic conditions (pH less than 3.0). Trivalent chromium can then be precipitated along with the other metals. Cyanide, which complexes with metals and keeps them in solution, can be oxidized with sodium hypochlorite or chlorine gas at a pH of 9.0 to 10.5 (alkaline chlorination) to free the metals so they can subsequently be removed by precipitation.

5.3 LEACHATE CONTAINING ORGANIC CONTAMINANTS

Leachate containing primarily organic contaminants can be treated effectively by stripping, adsorption, and/or biological treatment processes. Biological treatment processes are typically preceded by equalization and neutralization for protection of the microorganisms from toxic or inhibitory conditions and followed by sedimentation and/or filtration for separation of biological solids. For high-strength leachate, two biological units can be used in sequence (e.g., a trickling filter followed by an activated-sludge system), with the first serving as a roughing unit for partial degradation of the organics. Stripping and adsorption processes, on the other hand, are typically preceded by sedimentation and/or filtration for prevention of plugging of the packing material or granular activated carbon. The process flow diagram for the leachate pretreatment plant at the Love Canal site in Niagara Falls, New York (Figure 3) exemplifies removal of nonbiodegradable (refractory) organics from leachate. The most cost-effective treatment of leachate containing biodegradable and refractory organics includes a combination of biological and adsorption processes. Normally, biological treatment precedes carbon adsorption in the process train. With this arrangement, the biological operation substantially reduces the downstream organic loading on the carbon adsorbers. Biological degradation and carbon adsorption can also be performed in a single operation, as in the patented PACT process (see Subsection 6.3.3).

TABLE 3. COMPARATIVE COSTS OF LEACHATE TREATMENT TECHNOLOGIES (1986 dollars)

Technology	25 gal/min		50 gal/min		100 gal/min	
	Capital	Annual O&M	Capital	Annual O&M	Capital	Annual O&M
Equalization	86,000	6,300	126,000	9,400	178,000	11,400
Sedimentation	99,000	4,800	121,000	5,100	163,000	8,500
Granular-media filtration	17,000	4,900	25,000	6,200	38,000	7,400
Oil/water separation	57,000	1,700	63,000	1,900	75,000	2,300
Neutralization	28,000	3,100	49,000	3,500	53,000	4,000
Precipitation/flocculation/sedimentation	171,000	16,000	229,000	30,000	312,000	58,000
Oxidation/reduction	96,000	3,500	121,000	4,300	162,000	5,600
Carbon adsorption	65,000	38,000	107,000	58,000	163,000	112,000
Air stripping	72,000	38,000	118,000	70,000	208,000	137,000
Steam stripping	76,000	37,000	97,000	71,000	117,000	137,000
Reverse osmosis	69,000	30,000	126,000	41,000	235,000	76,000
Ultrafiltration	55,000	32,000	88,000	47,000	134,000	93,000
Ion exchange	59,000	10,000	83,000	16,000	118,000	26,000
Wet-air oxidation	421,000	82,000	567,000	154,000	826,000	269,000
Activated sludge	184,000	18,000	249,000	28,000	364,000	47,000
Sequencing batch reactor	109,000	10,000	158,000	16,000	224,000	26,000
Powdered activated carbon treatment (PACT)	249,000	39,000	340,000	73,000	492,000	138,000
Rotating biological contactor	103,000	13,000	183,000	20,000	383,000	36,000
Trickling filter	150,000	15,000	239,000	31,000	345,000	58,000
Chlorination	31,000	5,000	47,000	7,000	71,000	9,000

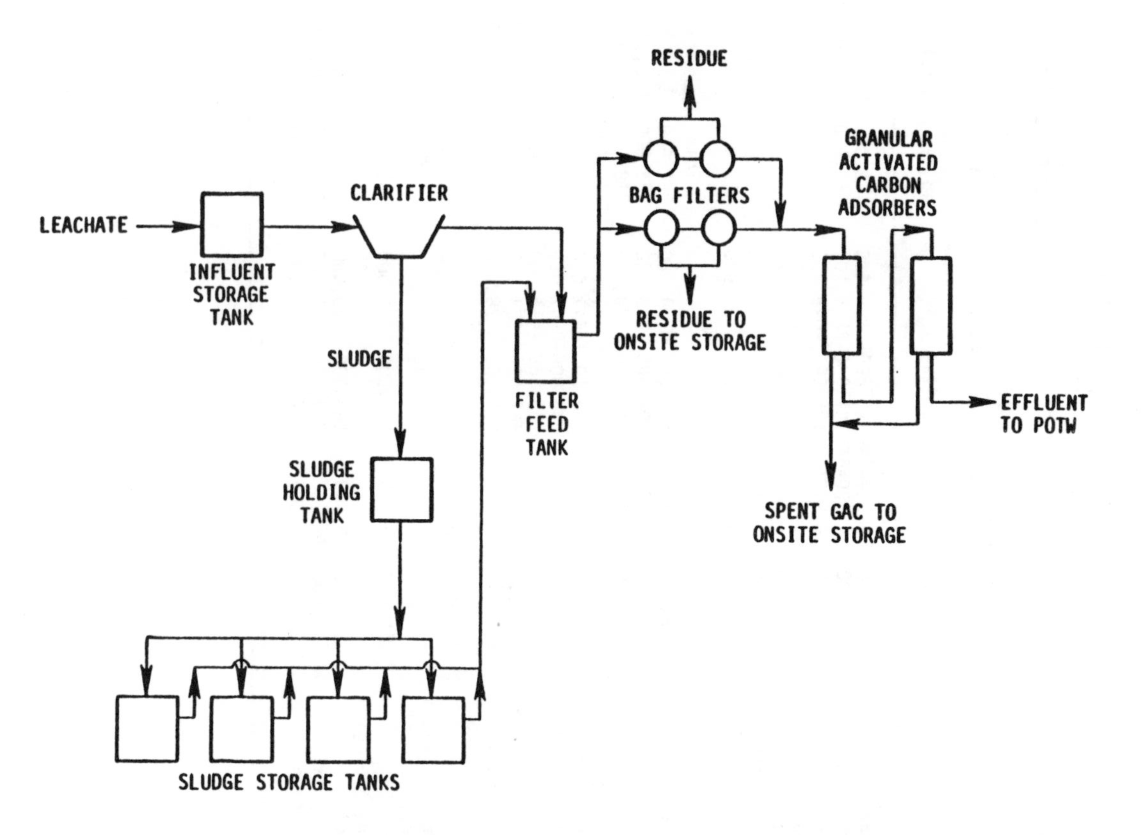

Figure 3. Flow diagram of the Love Canal leachate pretreatment plant.

5.4 LEACHATE CONTAINING INORGANIC AND ORGANIC CONTAMINANTS

In most cases, hazardous waste leachate contains both inorganic and organic contaminants, and the treatment trains required to treat these waste streams involve combinations of the process schemes described in Subsections 5.2 and 5.3. The best overall treatment efficiencies generally can be achieved by removing the inorganic constituents first and then removing the organic constituents. This approach protects the biological, adsorption, and stripping processes from problems caused by metals toxicity, corrosion, and scaling.

The Stringfellow leachate pretreatment plant in Glen Avon, California (Figure 4) illustrates a typical process train for leachate containing both inorganic and organic contaminants. Metals are removed from Stream A by precipitation/flocculation/sedimentation. The clarifier overflow is filtered and then mixed with Stream B. Organics are removed from the combined leachate stream by carbon adsorption, and the effluent is discharged to a POTW.

Table 4 summarizes the process trains that have been selected or proposed for treatment of hazardous waste leachate at Superfund sites. This table also contains case-study examples of process trains that incorporate innovative treatment technologies [e.g., sequencing batch reactor, powdered activated carbon treatment (PACT), and wet-air oxidation]. The information in this table was compiled from a review of approximately 130 Records of Decision available as of June 1986 and from responses to inquiries in each of the EPA Regions. A limited number of site visits were conducted to gather operating and performance data; these data are reported in Section 6.

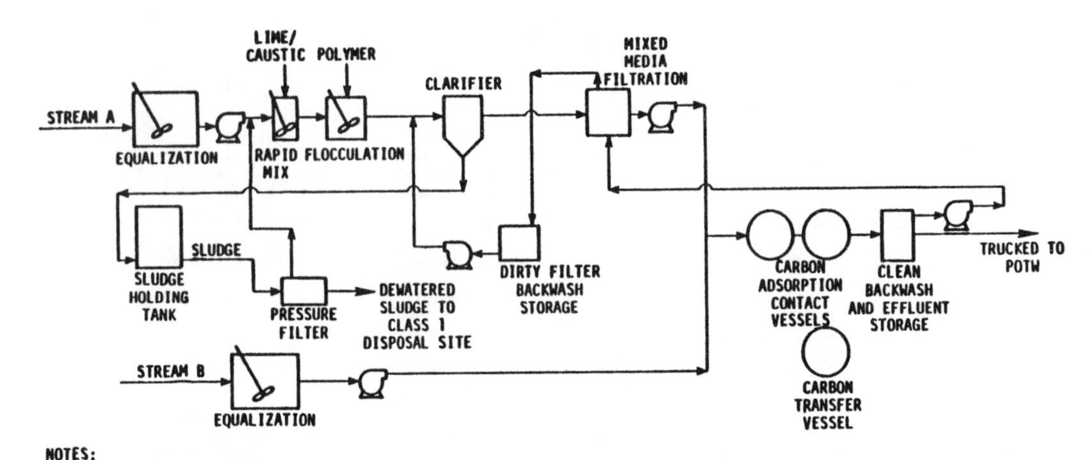

NOTES:

1. STREAM A IS FROM WELLS OW-1, OW-2, OW-4, IW-1 AND THE FRENCH DRAIN. AVERAGE FLOW IS EXPECTED TO BE 20 GPM. DESIGN FLOW IS 50 GPM.
2. STREAM B IS FROM MID-CANYON WELLS IW-2 AND IW-3. AVERAGE FLOW IS EXPECTED TO BE 40 GPM. DESIGN FLOW IS 80 GPM.
3. CURRENTLY (FEBRUARY 1986), INFLUENT IS TRUCKED FROM WELLS AND FRENCH DRAIN. IN THE NEAR FUTURE INFLUENT WILL BE PUMPED DIRECTLY INTO THE PLANT.

Figure 4. Flow diagram of the Stringfellow leachate pretreatment plant.

TABLE 4. LEACHATE TREATMENT CASE STUDY SITES

Site/location	Contaminants	Unit treatment processes	Discharge point	Source
Bofors-Nobel, Inc. Muskegon, Michigan	Dichloroethylene Orthochloroaniline Dichlorobenzidine	Neutralization Powdered activated carbon treatment/ wet-air oxidation	POTW	Meidl and Wilhelmi 1986
CECOS International, Inc. Niagara Falls, New York	Volatile organics Phenol	Equalization Neutralization Sequencing batch reactor Granular-media filtration Carbon adsorption	POTW	Staszak et al. undated
*Gloucester Environmental Management Services (GEMS) Landfill Gloucester Township, New Jersey	Volatile organics	Air stripping/vapor-phase carbon adsorption	POTW	EPA 1985b
*Helen Kramer Landfill Mantua Township, New Jersey	Heavy metals Volatile organics Phenols	Equalization Precipitation/flocculation/sedimentation Air stripping/vapor-phase carbon adsorption Activated sludge Granular-media filtration Carbon adsorption Chlorination	POTW or surface water	EPA 1985c
*Heleva Landfill North Whitehall Township, Pennsylvania	Heavy metals Volatile organics Dissolved organics	Precipitation/flocculation/sedimentation Neutralization Activated sludge Air stripping Carbon adsorption	Surface water	EPA 1985d
Hyde Park Landfill Niagara Falls, New York	Phenol HET acid Benzoic acid o-, m-, p-Chlorobenzoic acid	Equalization Neutralization/sedimentation Sequencing batch reactor Carbon adsorption	POTW	Ying et al. 1986
*Lipari Landfill Mantua Township, New Jersey	Heavy metals Volatile organics Phenols	Equalization Precipitation/flocculation/sedimentation Air stripping/vapor-phase carbon adsorption Granular-media filtration Carbon adsorption	POTW	EPA 1985e

(continued)

TABLE 4 (continued)

Site/location	Contaminants	Unit treatment processes	Discharge point	Source
*Love Canal Niagara Falls, New York	Volatile organics Semivolatile organics (acid extractables, base/neutral extractables) Dioxin	Equalization Sedimentation Bag filtration Carbon adsorption	POTW	Shuckrow, Pajak, and Touhill 1982
*New Lyme Landfill Ashtabula County, Ohio	Heavy metals Volatile organics Refractory organics	Neutralization/sedimentation Rotating biological contactor Precipitation/flocculation/sedimentation Carbon adsorption	Surface water	EPA 1985f
*Pollution Abatement Services (PAS) Site Oswego, New York	Heavy metals Volatile organics Semivolatile organics (acid extractables, base/neutral extractables)	Equalization Precipitation/flocculation/sedimentation Carbon adsorption Neutralization Granular-media filtration	Not specified	EPA 1984a Rothman, Gorton, and Sanford 1984
*Sand, Gravel, and Stone Site Elkton, Maryland	Heavy metals Volatile organics Semivolatile organics (acid extractables, base/neutral extractables)	Equalization Reduction Precipitation/flocculation/sedimentation/sludge dewatering Neutralization Filtration Carbon adsorption	Ground water/surface water	EPA 1985g
*Stringfellow Acid Pits Glen Avon, California	Heavy metals Organics	Equalization Precipitation/flocculation/sedimentation/sludge dewatering Granular-media filtration Carbon adsorption	POTW	EPA 1984b
*Sylvester Site (Gilson Road Site) Nashua, New Hampshire	Heavy metals Volatile organics Alcohols, ketones	Precipitation Neutralization Filtration High-temperature air stripping/fume incineration Activated sludge (extended aeration)	Ground water	EPA 1983
*Tyson's Dump Upper Merion Township, Pennsylvania	Volatile organics	Air stripping/vapor-phase carbon adsorption	Surface water	EPA 1984c

* NPL Superfund site.

6. Leachate Treatment Unit Processes

This section profiles 20 unit processes with demonstrated or potential applicability to the treatment of hazardous waste leachate. The technologies are classified as pretreatment operations, physical/chemical treatment operations, biological treatment operations, and post-treatment operations. The order in which the technologies within each category are presented reflects the reliability of the processes for leachate treatment applications (i.e., technologies that have been widely demonstrated are presented first; innovative technologies or technologies that have not been demonstrated with hazardous waste leachate are presented last).

The applicability of the profiled technologies to the treatment of hazardous waste leachate is based on a review of the 14 case-study sites presented in Table 4 (p. 21) or, where no experience exists, on the use of best engineering judgment. As the EPA and its contractors gain experience in this field, many of the existing information gaps will be filled (particularly those in the area of performance efficiency).

6.1 PRETREATMENT OPERATIONS

6.1.1 Equalization

Process Description--

Because leachate is subject to large fluctuations in volume and strength, operational problems can arise in leachate treatment plants when attempts are made to automate chemical additions or to maintain an active biomass for biological treatment processes. Equalization, which entails mixing the incoming leachate in a large tank or basin and discharging it to the treatment plant at a constant rate, improves the efficiency, reliability, and control of downstream processes by providing them with a more uniform feed.

Equalization tanks or basins may be designed as either in-line or side-line units, as illustrated in Figure 5. With in-line equalization, the entire daily flow passes through the basin, and leachate is discharged to the treatment plant at an essentially constant rate. With side-line equalization, only the flow above the average daily flow rate is diverted into the basin; when the flow rate falls below the daily average, leachate from the equalization basin is discharged to the treatment plant to bring the flow rate up to the average level. In-line equalization is the preferred arrangement for leachate treatment applications because contaminant concentrations as well as flow rates are equalized in this approach. Side-line systems typically only provide flow-rate equalization.

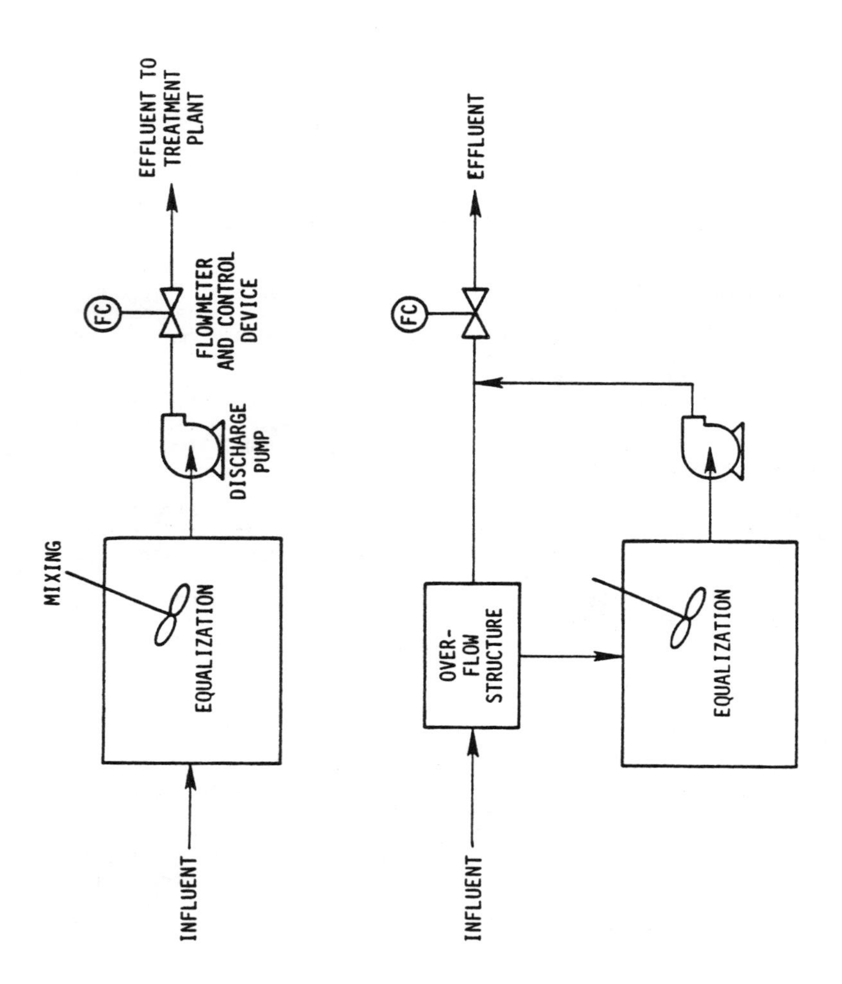

Figure 5. In-line and side-line flow equalization.

Equalization tanks or basins can be constructed of steel, concrete, or (lined) earthen materials. Provisions should be made for mixing of the leachate to prevent the deposition of solids. Discharge pumping and flow control are also required.

Applicability to Hazardous Waste Leachate--

Because the composition and volume of leachate vary greatly, equalization is required to achieve optimum performance of the treatment system. When placed ahead of chemical operations in the treatment process train, equalization improves chemical feed control and process reliability. When placed ahead of biological operations, equalization minimizes shock loadings, dilutes inhibitory substances, stabilizes pH, and improves secondary settling. In plants that operate on an intermittent schedule, equalization tanks/basins double as influent storage tanks. Eight of the 14 case-study sites presented in Table 4 (p. 21) include equalization as the first step in the treatment process train.

Few disadvantages are associated with equalization. Solids and oil and grease present in the leachate may tend to accumulate on the basin walls, but these materials can be removed by spraying the walls with water. Mixing of the tank/basin contents may strip highly volatile components from the leachate. The need for control of volatilized compounds must be determined on a site-specific basis.

Design Considerations--

Primary considerations in the design of equalization tanks/basins include the required storage volume, mixing equipment, and control devices for pumping and discharge flow rates.

The required storage volume can be determined from the average daily flow rate and the magnitude of inflow fluctuations. The tank or basin is generally designed with excess capacity to provide storage volume during periods of maintenance or regeneration of downstream processes.

Mixing requirements for leachate containing 200 mg/liter of suspended solids range from 0.02 to 0.04 hp per 1000 gal of storage (EPA 1977). Pumping will normally be required and may precede or follow equalization. Discharge from the tank/basin is regulated with a flow-control device and should be monitored with a flowmeter.

Because of the low flow rates and relatively high contaminant concentrations associated with hazardous waste leachate, above-ground tanks or concrete basins are generally preferred over lined surface impoundments.

Performance--

Equalization is generally reliable and can improve the performance of sensitive operations such as carbon adsorption, biological treatment, chemical precipitation, and ion exchange (EPA 1982a).

Costs--

Figure 6 presents capital and annual O&M costs of an equalization system, and Figure 7 presents a breakdown of these costs. The capital costs of equalization for streams of 25 to 100 gal/min range from $86,000 to $178,000;

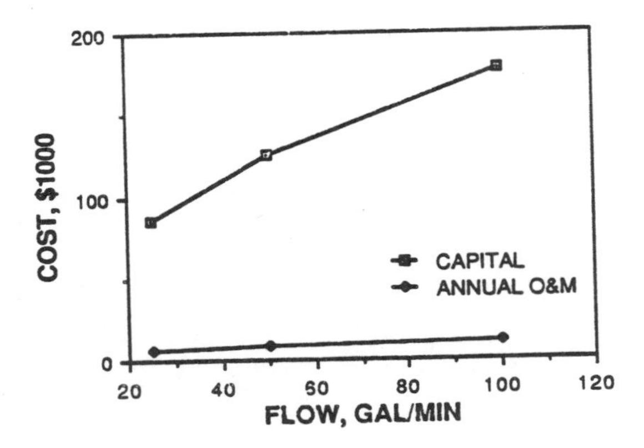

Figure 6. Capital and annual O&M costs of an equalization system.

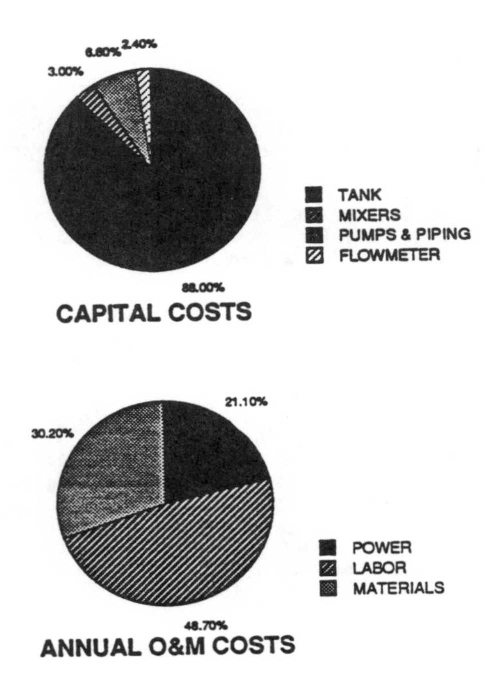

Figure 7. Breakdown of capital and annual O&M costs of an equalization system.

the tank or basin accounts for about 88 percent of the capital outlay. The annual O&M costs range from $6,300 to $11,400. The costs shown in this report are based on the use of an above-ground, carbon-steel storage tank with a retention time of 24 h (EPA 1982a; Peters and Timmerhaus 1980; Guthrie 1969; Barrett 1981; U.S. Army Corps of Engineers 1985; Richardson Engineering Services 1979).

Development of a minimum-cost equalization system requires the establishment of the minimum required retention time. The minimum retention time is determined by the expected variation in the feed to the treatment system and how much variation can be tolerated in the actual treatment steps. Sometimes downstream operating procedures or treatment methods can be adjusted to reduce equalization costs.

6.1.2 Sedimentation

Process Description--

Sedimentation is the gravitational settling of suspended particles that are heavier than water in a large tank or basin under quiescent conditions. The basin (called a clarifier) may be rectangular (Figure 8) or circular (Figure 9). In either case, the settled solids are collected mechanically on the bottom of the clarifier and pumped out as sludge underflow. Clarifiers are also equipped with surface-skimming equipment to remove floating scum. Chemical coagulants may be added to the clarifier to improve the settleability of fine particles and colloidal substances.

Applicability to Hazardous Waste Leachate--

Sedimentation is widely used for the removal of settleable solids and immiscible liquids, including oil and grease and some organics. Although hazardous waste leachate typically contains only small loadings of suspended solids, sedimentation may be included as a pretreatment step because of the sensitivity of many downstream processes to fouling and interference from suspended solids.

Frequently, sedimentation is included in leachate treatment process trains for separation of solids generated by chemical and biological processes. It is an integral part of every precipitation/flocculation process (Subsection 6.2.2), activated sludge process (Subsection 6.3.1), and powdered activated carbon treatment (PACT) process (Subsection 6.3.3).

The wet sludge underflow produced by sedimentation may be hazardous and requires further treatment and disposal. Sedimentation may also generate a nonaqueous organic liquid phase, which must be recovered and disposed of. Section 7 addresses management alternatives for sludge and concentrated liquid waste streams.

Design Considerations--

The design of clarifiers for removal of suspended solids is based on the settling rate or rise rate of the smallest particles to be removed, expressed as flow rate per unit area. Most applications fall within a range of 0.2 to 1.0 gal/min per ft^2 (300 to 1500 gal/day per ft^2) (Metcalf & Eddy 1985).

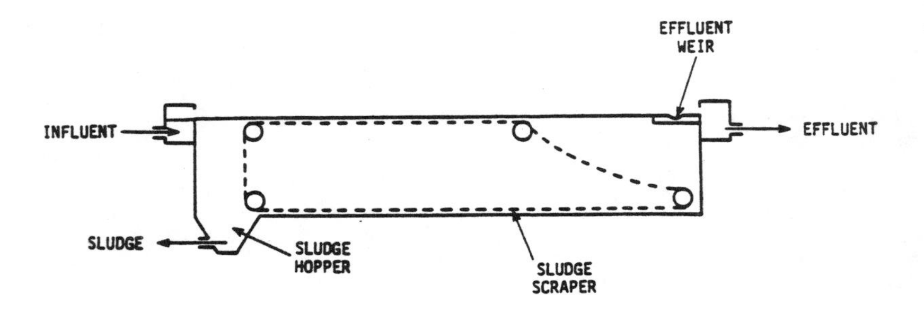

Figure 8. Schematic diagram of a rectangular clarifier.

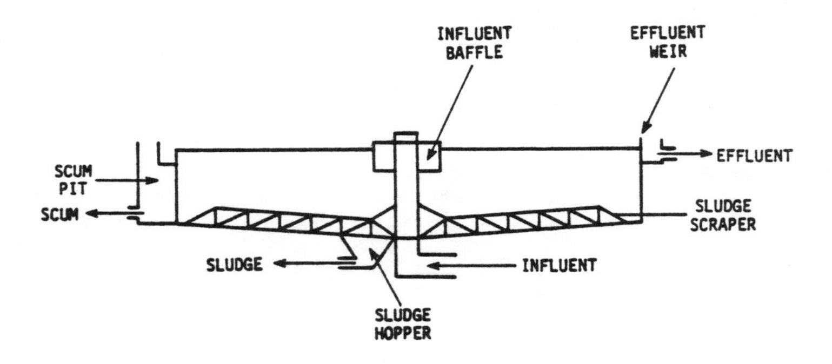

Figure 9. Schematic diagram of a circular clarifier.

Horizontal velocities must be limited to prevent scouring of settled solids from the sludge bed and their eventual escape in the effluent.

Performance--

Both circular and rectangular clarifiers are used widely and are considered highly reliable if properly operated and maintained. Common operational problems include plugging of sludge hoppers not equipped with cross collectors and rising sludge.

If designed and operated properly, pretreatment clarifiers can achieve removals of 50 to 65 percent total suspended solids, and they will generate an underflow sludge solids concentration of 3 to 7 percent (EPA 1980). Effluent suspended solids in the range of 20 to 50 mg/liter are typical (Metcalf & Eddy 1985).

Costs--

Figure 10 presents capital and annual O&M costs of a sedimentation basin, and Figure 11 presents a breakdown of these costs. The capital costs of sedimentation for 25- to 100-gal/min streams range from $99,000 to $163,000; the clarifier makes up about 61 percent of the capital outlay. The annual O&M costs range from $4,800 to $8,500; residuals management makes up about 39 percent of these expenses. The clarifier design assumes a hydraulic loading of 600 gal/day per ft^2 of surface (EPA 1982a; Richardson Engineering Services 1979; Cran 1981; Peters and Timmerhaus 1980; Barrett 1981).

6.1.3 Granular-Media Filtration

Process Description--

Filtration is a physical process whereby suspended solids are removed from leachate by forcing the fluid through a porous medium. A granular-media filter consists of a bed of granular material (typically sand or sand with anthracite or coal) contained within a vessel and supported by an underdrain system (Figure 12). As water laden with suspended solids passes through the bed of filter medium, the particles become trapped on top of and within the bed. This either reduces the filtration rate at a constant pressure or increases the amount of pressure needed to force the water through the filter. When the maximum pressure drop is reached or when breakthrough of the filter occurs (as indicated by elevated suspended solids in the effluent), the filter is backwashed at high velocity with treated effluent to dislodge the trapped particles. The backwash water, which contains high concentrations of solids, is typically recycled to the headworks of the treatment plant. Periodic backwashing of the filter medium is essential for continued operation of the filtration system.

Granular-media filtration systems may be classified by 1) direction of flow, 2) type of filter beds, 3) wastewater driving force, and 4) method of flow-rate control. Filtration systems may be operated in a downflow, upflow, or biflow mode. Downflow filters are used most commonly, but upflow filters are believed to be more efficient. Downflow filters may incorporate single-medium, dual-media, or tri-media filter beds. For high-turbidity leachate, the dual- and tri-media filter beds are more desirable than single-medium filter beds because they have greater solids storage capacity.

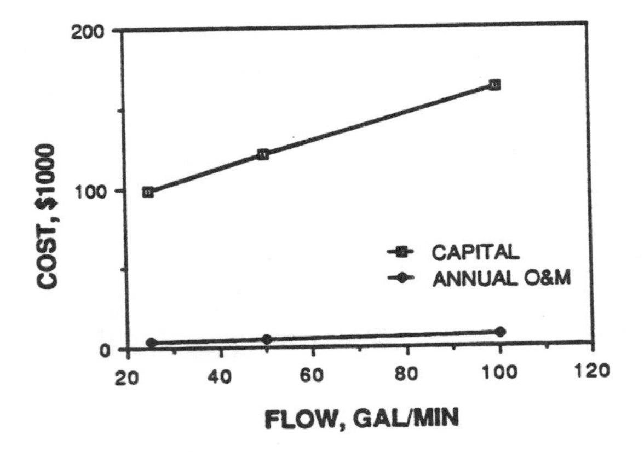

Figure 10. Capital and annual O&M costs of a sedimentation basin.

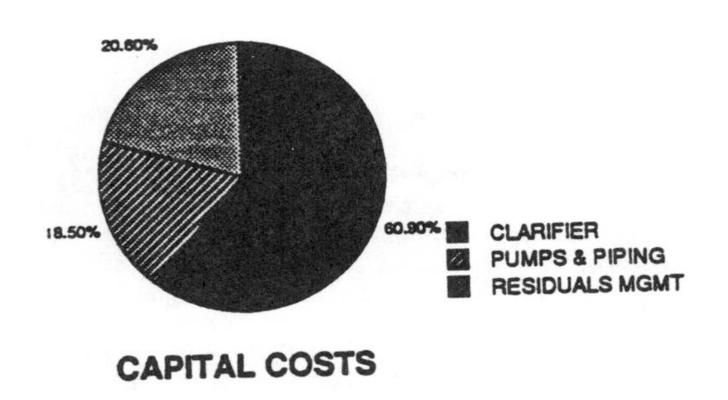

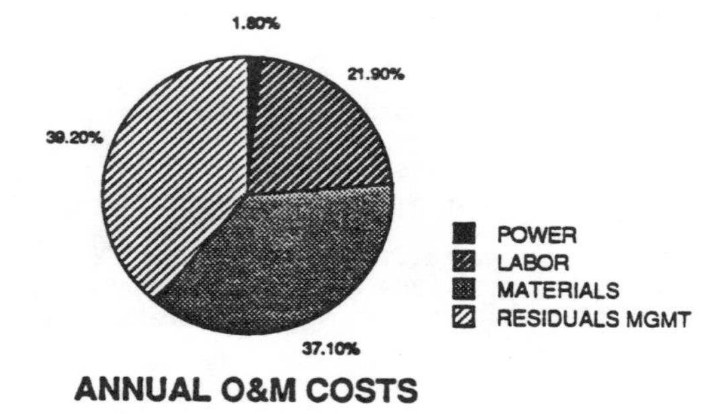

Figure 11. Breakdown of capital and annual O&M costs of a sedimentation basin.

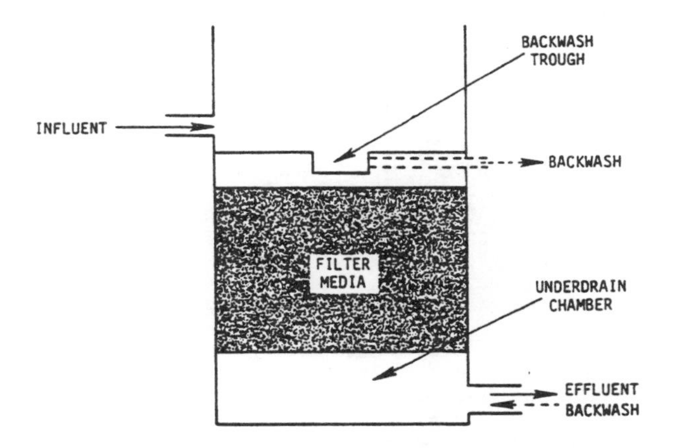

Figure 12. Schematic diagram of a downflow, granular-media gravity filter.

The driving force in the operation of granular-media filtration systems is either gravity or applied pressure. Gravity filter systems are used more commonly; however, pressure filters are widely used at smaller plants. The flow rate through gravity filters may be either at a constant rate or at a variable declining rate. In constant-rate filtration, the flow through the filter bed is maintained at a constant rate throughout the filter run by an effluent flow-control valve. In variable declining-rate filtration, the flow rate through the filter bed is initially high but decreases as head loss increases throughout the filter run. When the maximum allowable head loss is reached, the filter is taken off line for backwashing.

Applicability to Hazardous Waste Leachate--

Granular-media filtration is useful as a pretreatment step for adsorption processes (activated carbon), membrane separation processes (reverse osmosis, ultrafiltration), and ion exchange processes, which are rapidly plugged or fouled by high loadings of suspended solids. The most common application of granular-media filtration to hazardous waste leachate involves pretreatment prior to carbon adsorption. Five of the 10 leachate-treatment case-study sites presented in Table 4 (p. 21) that have been identified as using activated carbon adsorption include granular-media filters as pretreatment for removal of suspended solids.

Filtration may also be used as a polishing step after precipitation/flocculation or biological processes for removal of residual suspended solids in the clarifier effluent. In these applications, granular-media filtration should be preceded by gravity sedimentation of suspended solids to minimize premature plugging and backwashing requirements (Metcalf & Eddy 1985).

Design Considerations--

Key variables in the design of granular-media filter beds include the size of the filter medium, bed porosity (solids storage capacity), filter bed depth, filtration rate, maximum head loss, and influent leachate characteristics (Metcalf & Eddy 1979). Grain size of the medium will affect head loss and suspended solids removal. If the grain size is too small, the driving force used to push the leachate through the filter is spent overcoming the frictional resistance of the grain. If the grain size is too large, small particles will pass through the filter and into the effluent. Filter bed porosity or solids storage capacity determines the amount of solids that can be retained in the filter, and filter bed depth affects the head loss and the length of run. The filtration rate (which ranges from 2 to 15 gal/min per ft^2) affects the size of the filter, and the maximum head loss affects the length of the run. Influent leachate characteristics (e.g., suspended solids concentration, particle size and distribution, and floc strength) affect effluent quality.

Performance--

Granular-media filtration is used widely in the treatment of hazardous waste leachate because the system is simple and reliable. Granular-media filters can produce an effluent with a suspended solids concentration as low as 1 to 10 mg/liter (EPA 1985h). Such removal efficiencies will improve the performance of downstream processes.

Costs--

Figure 13 presents capital and annual O&M costs of a granular-media filter, and Figure 14 presents a breakdown of the O&M costs. The capital costs of granular-media filtration for 25- to 100-gal/min streams range from $17,000 to $38,000; because the capital costs are based on costs for a packaged unit, they are not broken down. The annual O&M costs range from $4,900 to $7,400. The unit used in this cost estimate is equipped with air scouring and a built-in backwash holding tank. Backwash frequency is controlled by the amount of solids in the feedwater; typically, the filters are designed to be backwashed about once a day. The filter in this estimate was designed for a filtering rate of 8 gal/min per ft^2 (Richardson Engineering Services 1979; EPA 1982a).

If continuous filtering is required, two filters must be installed so that one can operate while the other is in the backwash cycle. The cost numbers given here should be doubled if continuous operation is required.

6.1.4 Oil/Water Separation

Process Description--

Hazardous waste leachate that has not been diluted by ground water may be composed of two immiscible phases that will require separation before further treatment of the aqueous waste stream. Oil/water separation technology can be used to separate immiscible organics such as chlorinated solvents and PCB oils from leachate.

Gravity separators offer the most straightforward, effective means for phase separation. They generally consist of simple, readily available tanks that provide space in which the oil/water mixture can reside in relative calmness and separate by natural gravity forces.

Coalescing separators, which use baffles in the tank to promote oil droplet agglomeration, provide more effective separation and can be used in

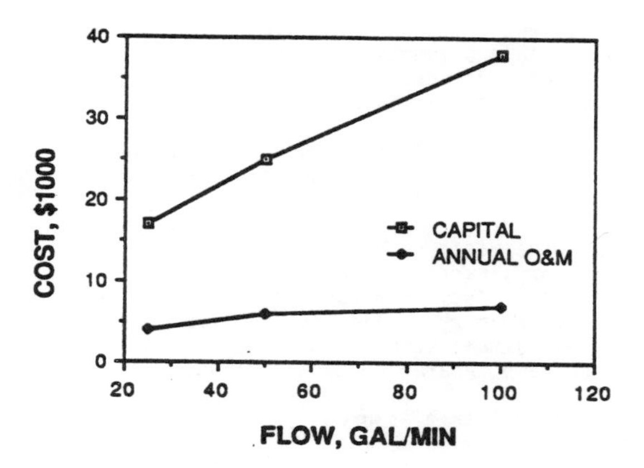

Figure 13. Capital and annual O&M costs of a granular-media filter.

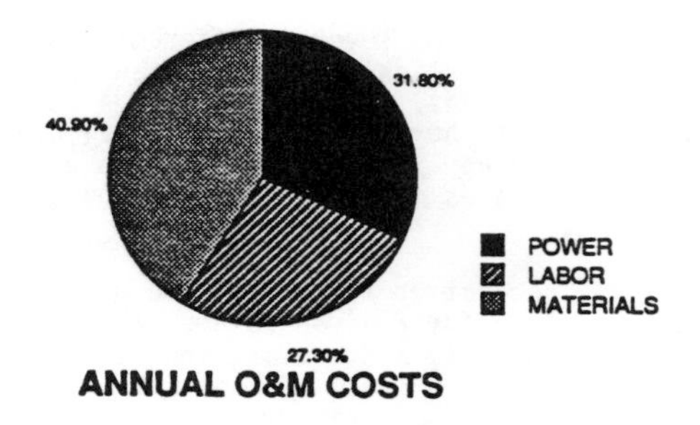

Figure 14. Breakdown of annual O&M costs of a granular-media filter.

situations where subsequent treatment processes cannot tolerate significant concentrations of immiscible organics. The agglomerated oil droplets form a continuous oil phase that more readily travels to the surface, where it can be easily skimmed off via a static pump or some other oil-removal mechanism.

Phase separation is frequently carried out in conjunction with sedimentation, as described in Subsection 6.1.2, and these processes may be accomplished concurrently in the same tank.

Applicability to Hazardous Waste Leachate--

The use of oil/water separation technology is limited to waste streams that are composed of two immiscible phases having significantly different specific gravities. Its primary application to hazardous waste leachate apperas to be as a pretreatment step ahead of easily fouled operations such as ion exchange or membrane separation technologies.

Leachate containing oil that is present as an emulsion will require the addition of an emulsion-breaking chemical for efficient treatment. Because contaminants in the leachate may affect the immiscibility of the oil and water phases, simple bench-scale tests also may be required to determine the separability of the liquid phases of the leachate.

The nonaqueous layer that is separated by this process is typically collected in an accumulator tank and then disposed of by incineration. In some cases, however, it may be recycled for product recovery or as an auxiliary fuel.

Design Considerations--

Phase separation involves several different equipment configurations, including horizontal and vertical cylinder decanters and cone-bottomed settlers for collecting settled solids. Figure 15 shows a more sophisticated coalescing oil/water separator. Both horizontal and vertical coalescing plate stacks may be included to cause any fine oil droplets to agglomerate into larger oil globules. Horizontal plates may be tilted at an angle to facilitate the sedimentation process.

The size of the separation tank should be determined by tests that measure the settling time for a given volume based on the leachate flow rate. Separators should be sized to accommodate the maximum flow rate expected.

Gravity feed provides the most effective delivery mechanism, as pumping can cause emulsification. Demulsifying agents can be added, however, to break emulsions and enhance separation.

Performance--

The efficiency of gravimetric oil/water separators is a function of oil concentration and droplet size, retention time, density difference between the two phases, and temperature. The surface area of the baffles also affects the efficiency of coalescing separators.

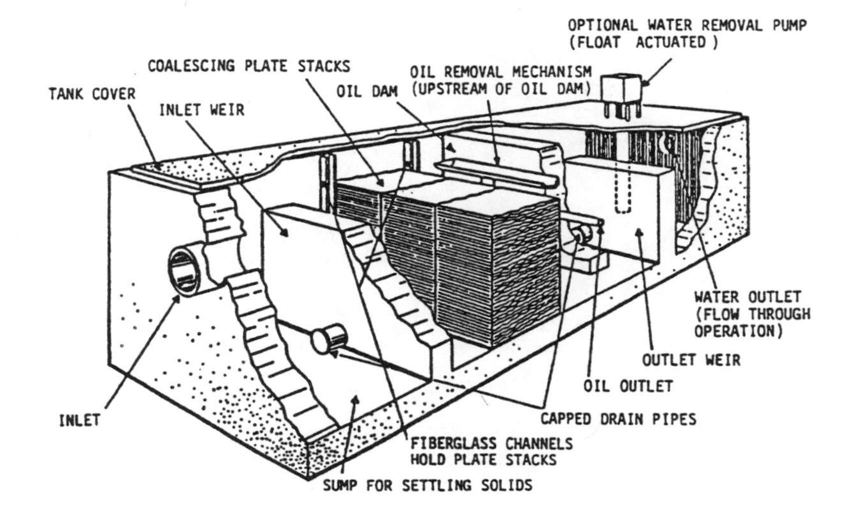

Figure 15. Coalescing oil/water separator design.

Source: Hansen and Rishel, undated.

Costs--

Figure 16 presents capital and annual O&M costs of a gravimetric oil/water separator, and Figure 17 presents a breakdown of the O&M costs. The capital costs of oil/water separation for 25- to 100-gal/min streams range from $57,000 to $75,000; because the capital costs are based on costs for a packaged unit, they are not broken down. The annual O&M costs range from $1,700 to $2,300. The nonaqueous layer is assumed to be recycled for product recovery (Richardson Engineering Services 1979).

6.2 PHYSICAL/CHEMICAL TREATMENT OPERATIONS

6.2.1 Neutralization

Process Description--

Neutralization of leachate exhibiting an extreme pH involves the addition of a base or an acid to the leachate to adjust its pH upward or downward, as required, to a final acceptable level (usually between 6.0 and 9.0). Bases commonly used for neutralization include lime (CaO), calcium hydroxide [$Ca(OH)_2$], caustic (NaOH), soda ash (NA_2CO_3), and ammonium hydroxide (NH_4OH); acids commonly used include sulfuric acid (H_2SO_4), hydrochloric acid (HCl), and nitric acid (NHO_3). Salts (soluble and insoluble) are formed as reaction byproducts of neutralization.

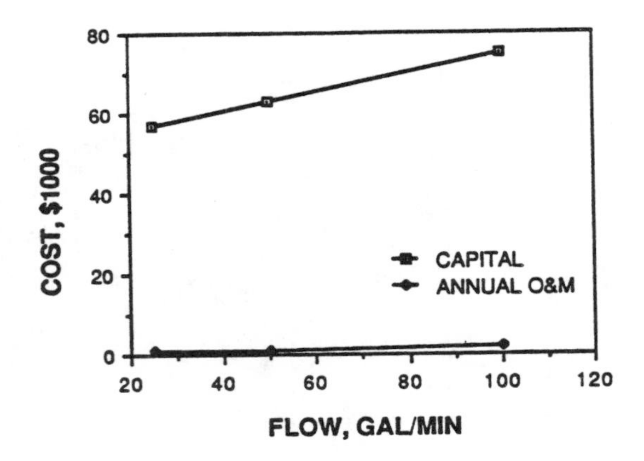

Figure 16. Capital and annual O&M costs of a gravimetric oil/water separator.

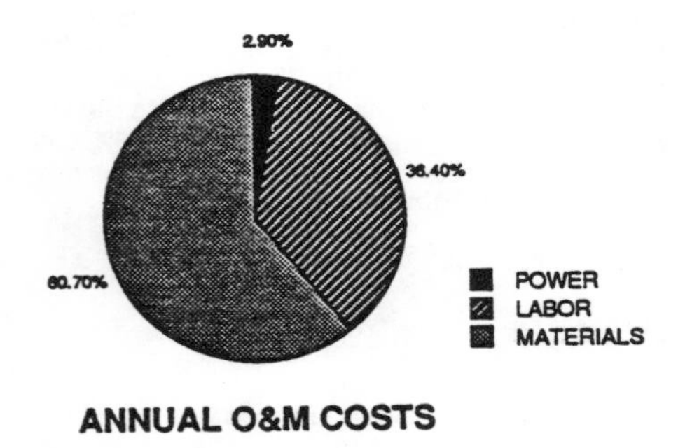

Figure 17. Breakdown of annual O&M costs of a gravimetric oil/water separator.

Typically, neutralization is carried out by complete mixing of the aqueous leachate with the neutralizing agent in a corrosion-resistant tank, as shown schematically in Figure 18. The tank may be operated in a batch or a continuous mode. Continuous-flow neutralization, which is generally suitable only for flow rates greater than about 70 gal/min, may be automatically controlled by using feedback, feed-forward, or multimode controllers (Patterson 1985).

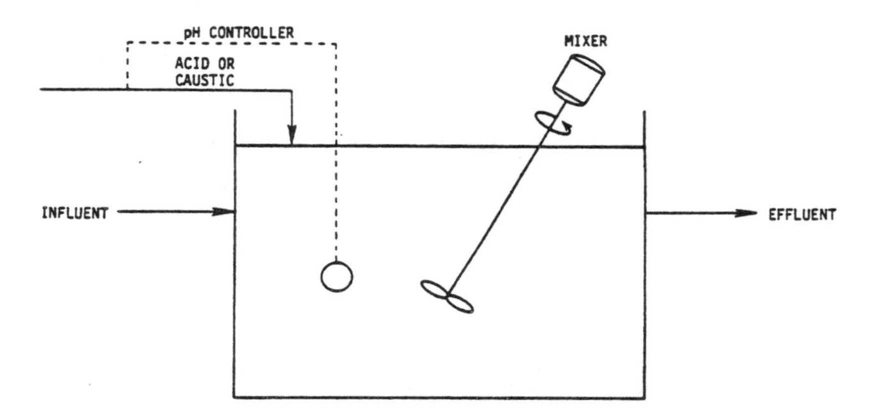

Figure 18. Schematic flow diagram of a neutralization system.

Applicability to Hazardous Waste Leachate--

Neutralization is one of the simplest and most common technologies available for the treatment of hazardous waste leachate. In most hazardous waste leachate treatment applications, neutralization serves as a form of pretreatment for optimization of the performance of pH-sensitive processes (particularly biological treatment processes) or for minimization of corrosion in more sophisticated physical/chemical treatments (especially membrane and stripping processes). In many of the leachate-treatment case studies reviewed, however, leachate is recovered in combination with large volumes of contaminated ground water, and this dilution has resulted in a pH that is acceptable for chemical or biological treatment without further neutralization.

Neutralization may also be applied as a post-treatment operation downstream of certain chemical processes that yield acidic or caustic effluents (e.g., oxidation/reduction). The use of post-treatment neutralization to meet final discharge criteria is particularly applicable where treated effluent is discharged to surface or ground water.

Equalization typically precedes neutralization in the treatment process train to dampen fluctuating influent conditions for improved process control.

Sedimentation and oil/water separation also may be included in the pretreatment system as needed. Treatment after neutralization (flocculation/sedimentation, filtration) is also frequently required to remove insoluble reaction byproducts (Metcalf & Eddy 1985).

Design Considerations--

Major process design considerations for neutralization systems include reagent selection and dosage, mixing, contact time, and process control. Factors that affect reagent selection include availability and cost of chemicals, ease of handling, reaction rate, reaction byproducts, and storage and equipment requirements (Patterson 1985). Sulfuric acid and various lime products are generally inexpensive; however, hydrochloric acid and sodium hydroxide may be more advantageous, as they often produce soluble reaction end products.

As with most processes for the treatment of hazardous waste leachate, special consideration must be given to the selection of vessels, piping, and instrumentation. Of particular concern is the selection of an automated pH control system capable of continued operation in the presence of concentrated chlorides (if HCl is used), scaling compounds (if lime-based chemicals are used), or suspended solids.

Performance--

Performance of neutralization systems is highly dependent on the reliability of automated control systems. The more simple and economical of these systems may not be able to achieve accurate pH control for fluctuating influent conditions; the more sophisticated and costly systems give more reliable performance (Patterson 1985).

Costs--

Figure 19 presents capital and annual O&M costs of a neutralization system, and Figure 20 presents a breakdown of the O&M costs. The capital costs of neutralization for 25- to 100-gal/min streams range from $28,000 to $53,000. Because the capital costs are based on costs for a packaged unit, they are not broken down. The annual O&M costs range from $3,100 to $4,000; chemicals make up about 32 percent of these expenses. The 25- and 50-gal/min systems include a reaction tank with a high-speed stirrer and a motorized PVC valve for caustic or acid addition. The 100-gal/min system consists of a reaction tank with a high-speed stirrer, two motorized PVC valves (one for rough and the other for fine pH adjustment), and two pH probes (one for the pH meter that controls pH and the other to record the pH) (Richardson Engineering Services 1979; EPA 1982a).

6.2.2 Precipitation/Flocculation/Sedimentation

Process Description--

Combined precipitation/flocculation/sedimentation is the most common method of removing soluble metals from leachate. Precipitation involves the addition of chemicals to the leachate to transform dissolved contaminants into insoluble precipitates. Flocculation promotes agglomeration of the precipitated particles, which facilitates their subsequent removal from the liquid phase by sedimentation (gravity settling) and/or filtration.

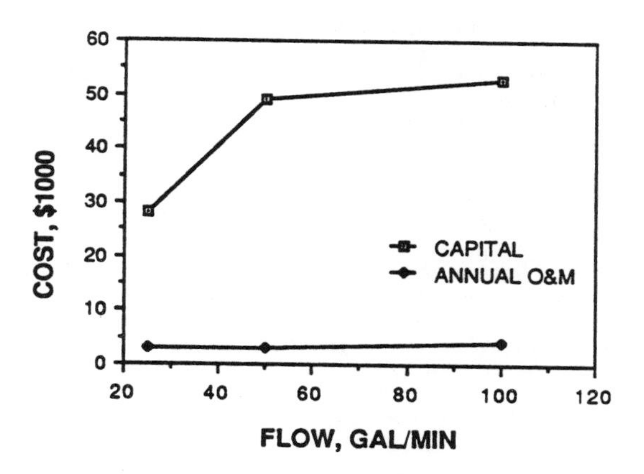

Figure 19. Capital and annual O&M costs of a neutralization system.

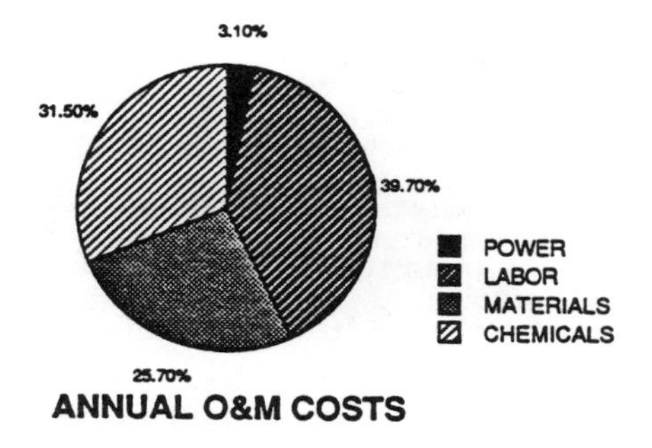

Figure 20. Breakdown of annual O&M costs of a neutralization system.

Metals can be precipitated from leachate as hydroxides, sulfides, or carbonates by adding an appropriate chemical precipitant and adjusting the pH to favor insolubility. Figure 21 shows the solubility of various metal salts as a function of pH. Although better removal efficiencies are possible with sulfide precipitation because of the low solubility of metal sulfides, hydroxide precipitation with lime or caustic as the precipitant is practiced more widely because of its materials-handling and cost advantages (Canter and Knox 1986).

The processes of precipitation, flocculation, and sedimentation can be carried out in separate basins, as shown schematically in Figure 22, or in a single basin (e.g., an upflow, solids-contact reactor-clarifier) with separate zones for each process. Precipitation requires rapid mixing to effect complete dispersion of the chemical precipitant, whereas flocculation requires slow and gentle mixing to promote particle contact. Frequently, flocculants such as alum, lime, ferric chloride, or polyelectrolytes are added along with the precipitant to reduce the repulsive forces between particles and bring about particle agglomeration and settling. Sedimentation (separation of the agglomerated solids from the liquid phase) is discussed in Subsection 6.1.2.

Applicability to Hazardous Waste Leachate--

Precipitation/flocculation/sedimentation is applicable to the removal of most metals [arsenic, cadmium, chromium (III), copper, iron, lead, mercury, nickel, and zinc] as well as suspended solids and some anionic species (phosphates, sulfates, and fluorides) from the aqueous phase of leachate (Shuckrow, Pajak, and Touhill 1982). All of the eight case-study sites with metals-contaminated leachate that are presented in Table 4 (p. 21) have designed or proposed treatment schemes incorporating precipitation for metals removal.

Because the fluctuating influent flow rate and metals content of leachate make chemical dosages difficult to control, equalization should be provided prior to precipitation. Also, nonaqueous liquids, including oils and miscible organics, should be removed during pretreatment.

Metals-complexing agents, including cyanide and ammonia, will inhibit chemical precipitation by forming soluble metal complexes. Removal of these compounds by oxidation/reduction or other processes prior to precipitation greatly improves treatment system performance. Such process interferences should be evaluated during treatability studies.

Because precipitation of most metals is conducted at an elevated pH, neutralization of the effluent may be required, particularly if a pH-sensitive biological treatment unit is included downstream. Precipitation/flocculation/-sedimentation generates large amounts of wet sludge that is likely to be considered hazardous because of its metals content. The subsequent treatment and disposal of this sludge is addressed in Subsection 7.1 of this report.

Design Considerations--

The design of precipitation/flocculation/sedimentation systems involves determination of required chemical dosages, optimum operating pH, degree of precipitation, reaction times, and sludge production and settling rates

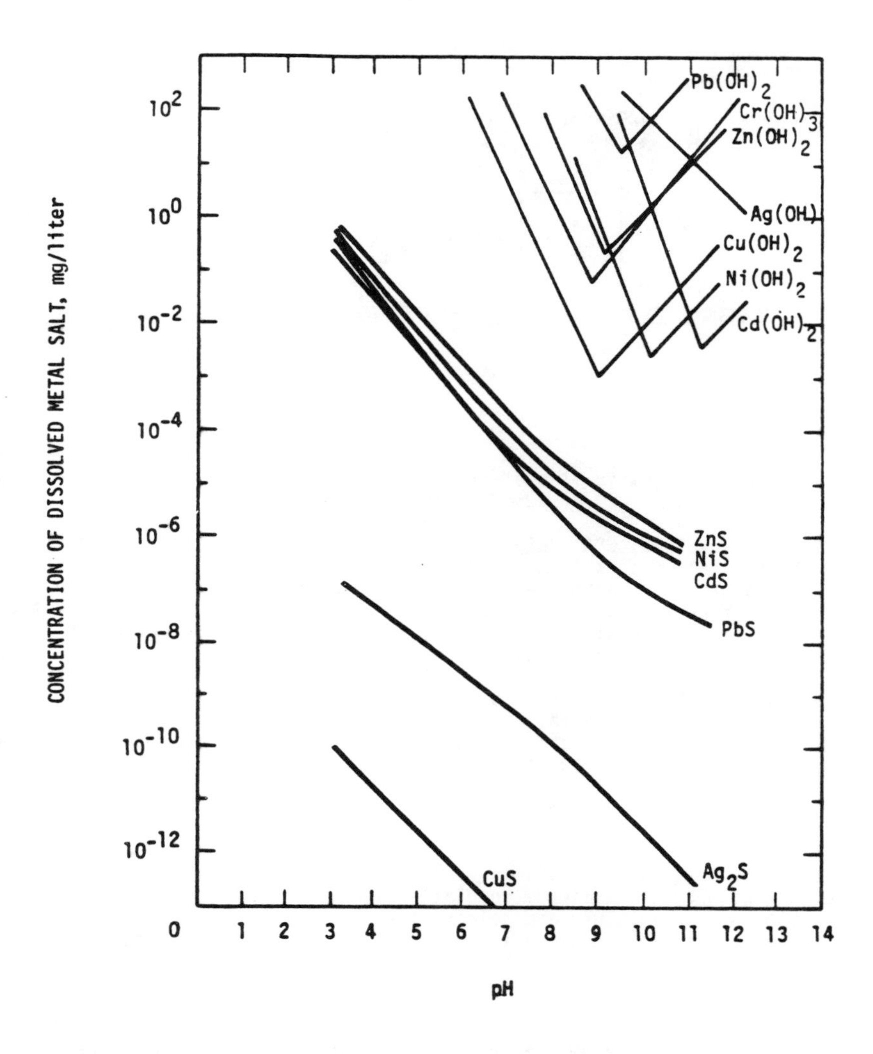

Figure 21. Solubility of metal hydroxides and sulfides as a function of pH.

Source: EPA 1985h.

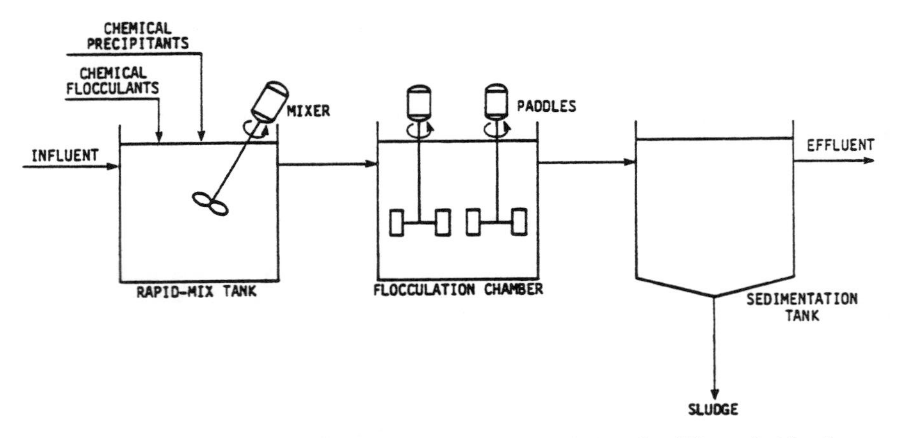

Figure 22. Schematic flow diagram of a precipitation/flocculation/-sedimentation system.

(Canter and Knox 1986; EPA 1982a). These parameters can be determined from simple bench-scale treatability studies (jar tests) for the chemicals of choice. Hydroxide or carbonate precipitation is preferable to sulfide precipitation for hazardous waste leachate applications because of the potential for the latter to generate toxic hydrogen sulfide gas. Depending on the nature of the leachate, chemical dosages can be quite high.

Because of the corrosive nature of many of the chemicals used for precipitation, special consideration should be given to the materials of construction of tanks, pumps, and piping. Packaged plants are available for low flow rates (10,000 to 2×10^6 gal/day), but these may require extensive modifications to handle the high level of precipitated solids that are characteristic of hazardous waste leachates.

Performance--

Effluent metal concentrations of less than 1 mg/liter are theoretically achievable with precipitation/flocculation/sedimentation. In practice, however, theoretical values are seldom attained because of the influence of complexing agents, fluctuations in pH, slow reaction rates, and poor separation of colloidal precipitates (Metcalf & Eddy 1985; Patterson 1985).

Costs--

Figure 23 presents capital and annual O&M costs of a precipitation/flocculation/sedimentation system, and Figure 24 presents a breakdown of these costs. The capital costs of precipitation/flocculation/sedimentation for 25- to 100-gal/min streams range from $171,000 to $312,000; the flocculator/clarifier accounts for about 28 percent of the capital outlay. The annual O&M costs range from $16,000 to $58,000; residuals management accounts for about 76 percent of these expenses (Richardson Engineering Services 1979; R.S. Means Co. 1986; EPA 1982a; Chemical Marketing Reporter 1986).

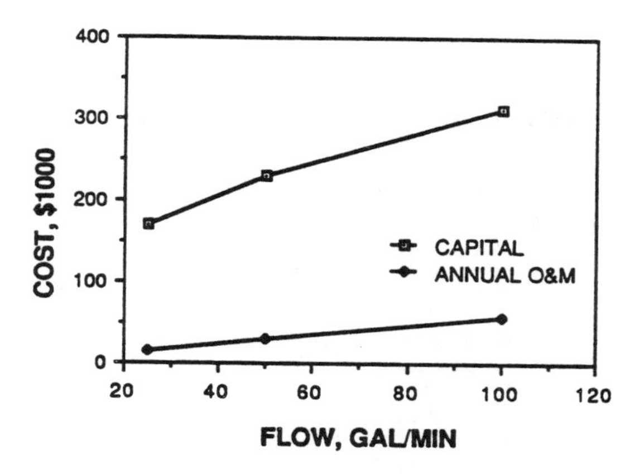

Figure 23. Capital and annual O&M costs of a precipitation/flocculation/sedimentation system.

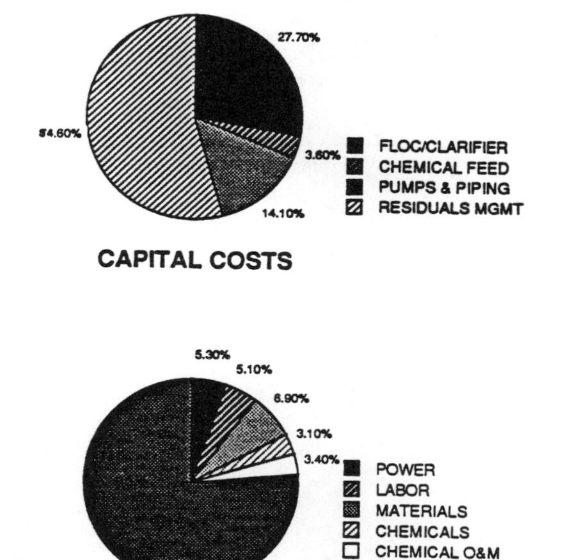

Figure 24. Breakdown of capital and annual O&M costs of a precipitation/flocculation/sedimentation system.

6.2.3 Oxidation/Reduction

Process Description--

Oxidation/reduction reactions are those in which the valence state of one reactant is raised while that of another is lowered. For example, in the reduction of hexavalent chromium to trivalent chromium by the use of sulfur dioxide, the oxidation state of chromium is reduced from +6 to +3 while that of sulfur is raised from +4 to +6, as given below:

$$\underset{(+6)}{2H_2CrO_4} + \underset{(+4)}{3SO_2} \rightleftarrows \underset{(+3)(+6)}{Cr_2(SO_4)_3} + 2H_2O$$

(oxidation: SO_2 (+4) → (SO_4) (+6); reduction: CrO_4 (+6) → Cr_2 (+3))

Oxidation/reduction of certain leachate constituents may render them non-hazardous or more amenable to removal by subsequent processes (e.g., precipitation, ion exchange, or biological treatment).

Oxidation/reduction involves the addition of a chemical oxidizing or reducing agent to leachate under a controlled pH. Common oxidizing agents include chlorine gas (plus caustic), calcium and sodium hypochlorite, chlorine dioxide, potassium permanganate, hydrogen peroxide, and ozone (alone or in combination with ultraviolent radiation); common reducing agents include sulfur dioxide, sodium sulfite salts (sodium bisulfite, metabisulfite, and hydrosulfite), sodium borohydrite, and the base metals (iron, aluminum, and zinc). Oxidation/reduction reactions (batch or continuous) are typically carried out in enclosed cylindrical vessels equipped with rapid-mix agitators (Figure 25). Reaction progress is monitored with an oxidation/reduction potential (ORP) probe.

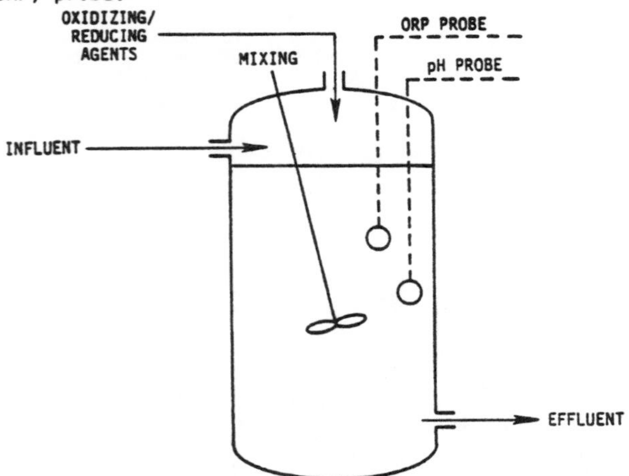

Figure 25. Schematic diagram of an oxidation/reduction reactor.

Applications/Limitations--

Oxidation/reduction technology is well developed and has many applications in industrial wastewater treatment. Oxidizable leachate constituents include organics (acids, aldehydes, mercaptans, phenols, polynuclear aromatic hydrocarbons, pesticides, PCB's, and other halogenated organics), cyanides, ammonia, and some metals (iron, manganese, and selenium). Reducible leachate constituents include a variety of metals (chromium, mercury, lead, silver, nickel, copper, and zinc). Metal-cyanide complexes can be treated by first oxidizing the cyanide and then reducing the metal (Metcalf & Eddy 1985; Patterson 1985).

The most common applications of oxidation/reduction to hazardous waste leachate include cyanide destruction and the reduction of hexavalent chromium to the less hazardous trivalent form. Only one oxidation/reduction operation (chromium reduction at the Sand, Gravel, and Stone site in Elkton, Maryland) was identified among the leachate-treatment case-study sites presented in Table 4 (p. 21).

Pretreatment requirements for oxidation/reduction operations are minimal and typically include only equalization and sedimentation. Oxidation/reduction of metals is usually followed by chemical precipitation/sedimentation, which produces a wet sludge. In such cases, the oxidation/reduction step is included to facilitate precipitation or to generate a less hazardous sludge. Partial oxidation with chlorine may result in the formation of toxic and/or odorous chlorinated organic species (Metcalf & Eddy 1985). Residual chlorine, ozone, or sulfites in the effluent may be damaging to downstream processes and should be neutralized.

Design Considerations--

The primary design considerations associated with oxidation/reduction systems include determination of the appropriate type and dosage of chemical reagent, the minimum contact time required to assure complete reaction, and the optimum solution pH (Patterson 1985). These parameters can be determined from laboratory bench studies.

Selection of the appropriate oxidizing/reducing agent is influenced by chemical and equipment costs, ease of handling, and safety. Chlorine gas, for example, is inexpensive, but dangerous to store and use; hypochlorite is twice as expensive, but much simpler and safer to use (Patterson 1985).

Oxidation/reduction reactions are typically exothermic and can be violent. For this reason, they are normally conducted at dilute concentrations and are frequently designed as batch processes. Chemical oxidizing and reducing agents are nonselective; consequently, leachate containing multiple oxidizable/reducible constituents will exert a higher chemical demand than will a less complex waste stream (Patterson 1985).

Performance--

The effectiveness of oxidation/reduction for a given constituent is directly related to the time of reaction and the degree to which interfering or competing constituents are present. In general, oxidation with ozone is much more rapid than oxidation with chlorine. For example, oxidation of cyanide to cyanate requires 10 to 15 min with ozone, compared with 0.5 to 2 h with chlorine (Patterson 1985). Reduction of chromium and other metals is more than 90 percent complete within 1 or 2 h (Metcalf & Eddy 1985).

Costs--

Figure 26 presents capital and annual O&M costs of an oxidation/reduction system, and Figure 27 presents a breakdown of these costs. The capital costs of oxidation/reduction for 25- to 100-gal/min streams range from $96,000 to $162,000; the reaction vessel accounts for about 60 percent of the capital outlay. The annual O&M costs range from $3,500 to $5,600; chemicals account for about 21 percent of these expenses. The reaction vessel is designed for a retention time of 2 h. Estimates of chemical costs for the system were based on the use of sodium bisulfite as a reducing agent (EPA 1982a; Peters and Timmerhaus 1980; Guthrie 1969; Barrett 1981; Chemical Marketing Reporter 1986).

6.2.4 Carbon Adsorption

Process Description--

Carbon adsorption is a separation technique for removing dissolved organics from leachate. The process involves passing the leachate through beds of granular activated carbon. Activated carbon, which has been specially processed to develop internal porosity, is characterized by a large specific surface area (300 to 2500 m^2/g). Contaminants are adsorbed from the leachate onto the carbon surface and held there by physical and chemical forces. Because the adsorption forces are relatively weak, the carbon surface can be regenerated.

The various means of contacting leachate with granular carbon include fixed-bed, expanded-bed, and moving-bed columns (illustrated schematically in Figure 28). In the fixed-bed adsorber, leachate is distributed at the top of the column, flows downward through the carbon bed (which is supported by an underdrain system), and is withdrawn at the bottom. When the pressure drop through the column becomes excessive (from the accumulation of suspended solids), the column is taken off line and backwashed with the treated effluent; the backwash water is then returned to the headworks of the plant for treatment. In the expanded-bed adsorber, leachate is introduced at the bottom of the column and flows upward through the bed at a velocity sufficient to suspend the carbon. Backwashing is not required because suspended solids pass through the bed with the effluent. In the moving-bed adsorber, leachate is introduced at the bottom of the column and flows upward through the carbon bed. Spent carbon is withdrawn intermittently from the bottom of the column and replaced with fresh carbon at the top; this, in effect, creates a countercurrent flow of carbon and leachate.

Activated carbon has a fixed adsorptive capacity. Breakthrough occurs when this capacity is approached, as indicated by elevated concentrations of organics in the adsorber effluent. Because of this breakthrough phenomenon, two columns are usually operated in series and a third ready to come on line when one of the columns is exhausted (Figure 29). The spent carbon may then be regenerated on site, returned to the supplier for regeneration, or disposed of offsite.

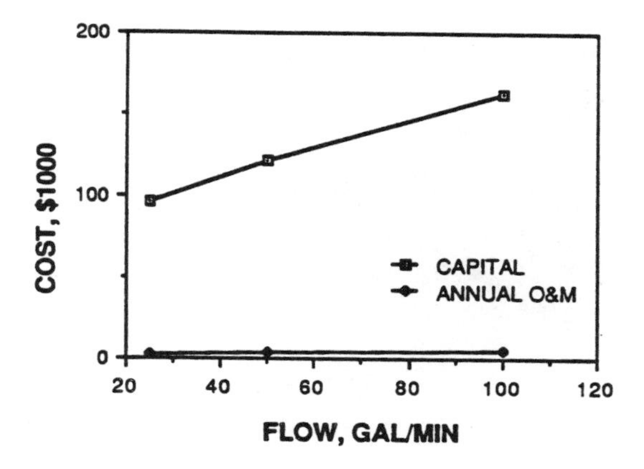

Figure 26. Capital and annual O&M costs of an oxidation/reduction system.

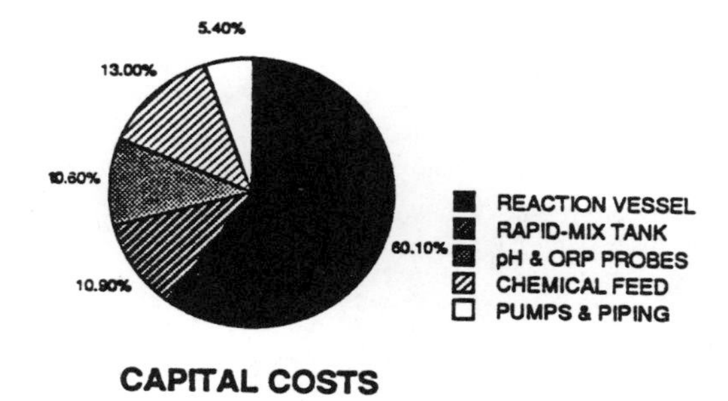

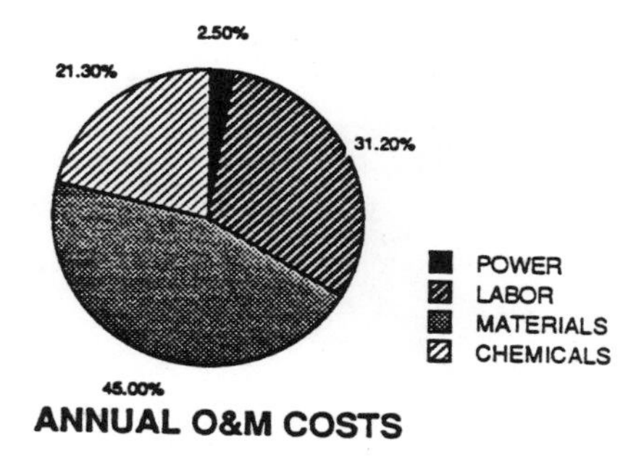

Figure 27. Breakdown of capital and annual O&M costs of an oxidation/reduction system.

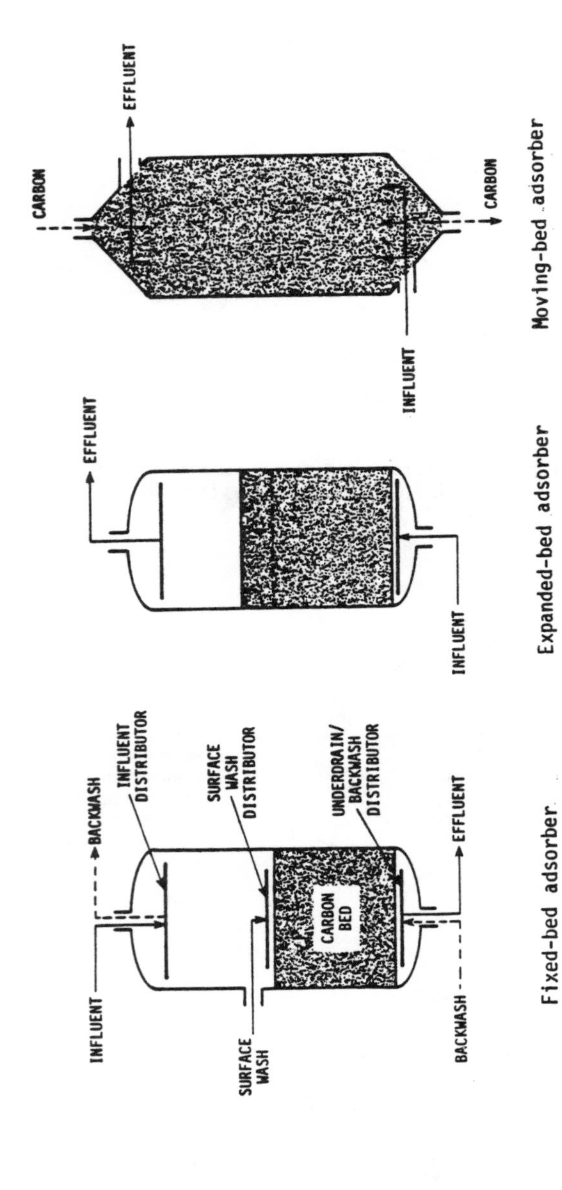

Fixed-bed adsorber Expanded-bed adsorber Moving-bed adsorber

Figure 28. Schematic diagrams of activated carbon adsorbers.

Figure 29. Two-stage carbon adsorption system with backup carbon adsorber.

Applicability to Hazardous Waste Leachate--

Granular activated carbon adsorption is a well-developed process that has become recognized as standard technology for the treatment of most hazardous waste leachates. It is especially well suited for the removal of mixed organic contaminants, including volatile organics, phenols, pesticides, PCB's, and foaming agents (EPA 1982a; Soffel 1978). Ten of the 14 leachate-treatment case-study sites presented in Table 4 (p. 21) use carbon adsorption for removal of various organic contaminants.

Carbon adsorption is economically competitive with air stripping for the removal of relatively low concentrations of volatile organics when VOC air emissions must be controlled. For higher contaminant loadings, carbon adsorption typically is used for effluent polishing of nonvolatile organics following air stripping.

Although carbon adsorption effectively removes some inorganics (including arsenic, cyanide, and chromium), hazardous waste leachate treatment facilities typically remove these contaminants through precipitation and/or oxidation/-reduction units upstream of the adsorber.

Effective pretreatment of leachate is critical to the successful operation of activated carbon adsorption units. If not removed in pretreatment, suspended solids and oil and grease will accumulate on the surface and in the first few inches of carbon. This blinds the adsorber and greatly increases the pressure drop across the filter bed. Granular-media filtration is usually provided upstream to minimize suspended solids loadings and to reduce the frequency of backwashing operations. In general, influent concentrations should be limited to 50 mg/liter suspended solids and 10 mg/liter oil and grease (EPA 1982a; Metry and Cross 1976). Acidification of the adsorber feed also may be required to prevent scaling.

Management alternatives for spent carbon are addressed in Subsection 7.4. Because of the high cost of activated carbon, regeneration is the most desirable alternative. Thermal regeneration will completely destroy compounds that decompose at temperatures of 600° to 1200°F; however, the presence of certain contaminants (e.g., PCB's) that are not destroyed at these temperatures, may render regeneration impractical and mandate incineration or disposal of the carbon instead.

Design Considerations--

Factors that affect the dynamics of the adsorption process include characteristics of the adsorbent (surface area, pore structure and size distribution, particle size, surface polarity), characteristics of the solute (solubility, molecular structure, molecular size, ionization, polarity), and characteristics of the aqueous system (temperature, pH, competing solutes, dissolved solids) (Patterson 1985).

The capacity of carbon for removing a particular compound from leachate can be approximated from an adsorption isotherm, which expresses the quantity of material that can be adsorbed per unit weight of carbon as a function of the equilibrium solute concentration at a constant temperature. In general, adsorption isotherms give a useful approximation of the treatability of leachate constituents, but they underestimate the carbon dosage required (EPA 1982a). Therefore, pilot tests should be performed to determine the design hydraulic load, bed depth, and contact time. Table 5 presents typical ranges for these parameters.

TABLE 5. OPERATING PARAMETERS FOR CARBON ADSORPTION[a]

Parameter	Operating range
Carbon dosage, lb COD/lb carbon	0.2 to 0.8
Hydraulic load, gal/min per ft^2	2 to 15
Bed depth, ft	10 to 30
Contact time, min	10 to 50

[a] Source: EPA 1982a.

Backwashing is normally provided for downflow, fixed-bed adsorbers, which are suspectible to blinding. Backwashing requirements can be minimized, however, through effective pretreatment (as described previously) and the use of low-pressure systems that limit compaction of the carbon beds.

Performance--

Carbon adsorption systems usually can be designed to effect greater than 99 percent removal of most organic contaminants. Because of the complex nature of hazardous waste leachate and the nonselectivity of carbon for specific hazardous constituents, however, effluent concentrations of target contaminants in the parts-per-billion range are difficult to achieve.

Carbon adsorption system reliability is largely a function of the pretreatment provided. In general, carbon filters that are provided with a low-turbidity, nonscaling feed can be expected to be highly reliable and require only infrequent backwashing.

Costs--

Figure 30 presents capital and annual O&M costs of a carbon adsorption system, and Figure 31 presents a breakdown of these costs. The capital costs of carbon adsorption for 25- to 100-gal/min streams range from $65,000 to $163,000; the pressure vessel and internals make up about 42 percent of the capital outlay. The annual O&M costs range from $38,000 to $112,000; replacement carbon makes up about 47 percent of these expenses and residuals management, 4 percent. The design of the system assumes a carbon-bed contact time of 30 min and a hydraulic loading of 5 gal/min per ft^2. Spent carbon is assumed to be disposed of offsite (EPA 1982a; DeWolf, Hearn, and Storm 1982; Barrett 1981; Peters and Timmerhaus 1980; personal communication from J. Jacobs, Calgon Carbon Corp., Houston, Texas, May 27, 1986).

6.2.5 Air Stripping

Process Description--

Air stripping is a mass-transfer process that uses air to remove volatile organic compounds from leachate. Types of air stripping devices include diffused aerators, mechanical surface aerators, coke tray aerators, sprays and spray towers, and packed towers (countercurrent and cross-flow). Countercurrent packed towers are best suited for leachate treatment applications because 1) they provide the greatest gas-liquid interfacial area for mass transfer, 2) they can be operated at higher air-to-water volume ratios than the other devices, and 3) emissions to the atmosphere are more easily controlled (Canter and Knox 1986).

A countercurrent packed tower (shown in Figure 32 and illustrated schematically in Figure 33) consists of a cylindrical shell containing randomly dumped packing on a support plate. Leachate is distributed uniformly on top of the packing with sprays or distribution trays and flows downward by gravity. The liquid can be redistributed at regular intervals to prevent channeling of the flow along the wall. Air is blown upward through the packing by forced or induced draft and flows countercurrently to the descending liquid. The volatile organics stripped from the leachate by the rising air are discharged to the atmosphere through the top of the column; effluent is discharged from the bottom of the column.

Applicability to Hazardous Waste Leachate--

Air stripping can effectively treat leachate containing organics that are volatile and only slightly water-soluble. As in the case of carbon adsorption, this technology has been widely demonstrated at Superfund sites; six of the leachate-treatment case-study sites presented in Table 4 (p. 21) have specified air stripping for removal of volatile organics.

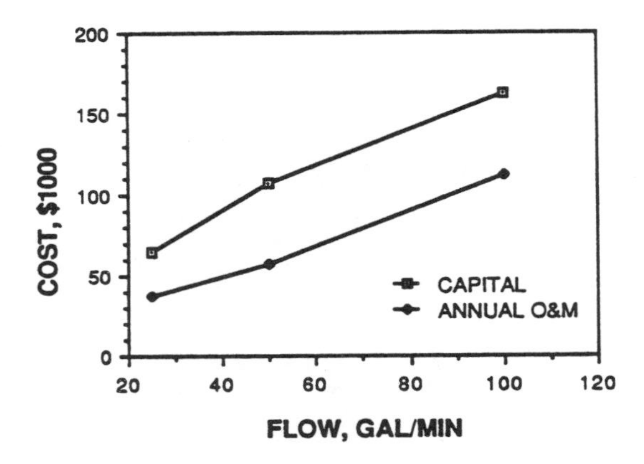

Figure 30. Capital and annual O&M costs of a carbon adsorption system.

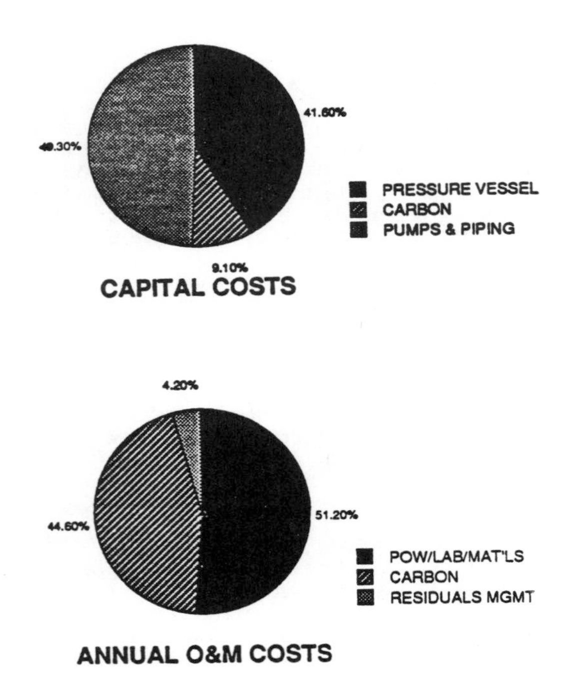

Figure 31. Breakdown of capital and annual O&M costs of a carbon adsorption system.

Figure 32. Packed-tower air stripper.

The applicability of air stripping for removal of a particular contaminant can be predicted by the use of vapor/liquid equilibria data. Because the vapor/liquid equilibrium behavior of a compound varies with temperature and the presence of other constituents, air-stripping efficiency should be determined experimentally in laboratory evaluations with actual leachate (Warner, Cohen, and Ireland 1980). Anticipated extremes of temperature and contaminant loading should be included in the treatability program.

High-temperature air stripping, in which the feed is preheated, has been applied to remove some chemicals (e.g., methyl ethyl ketone) that are not easily stripped at ambient temperatures (Johnson, Lenzo, and Sullivan 1985).

Pretreatment requirements for air stripping include removal of suspended solids and separation of nonaqueous phases. In systems where chemical neutralization or precipitation/sedimentation precede the stripper, sodium-based reactants should be selected over calcium-based reactants, as the latter can lead to scale formation. Acidification of the stripper feed can also help prevent scaling.

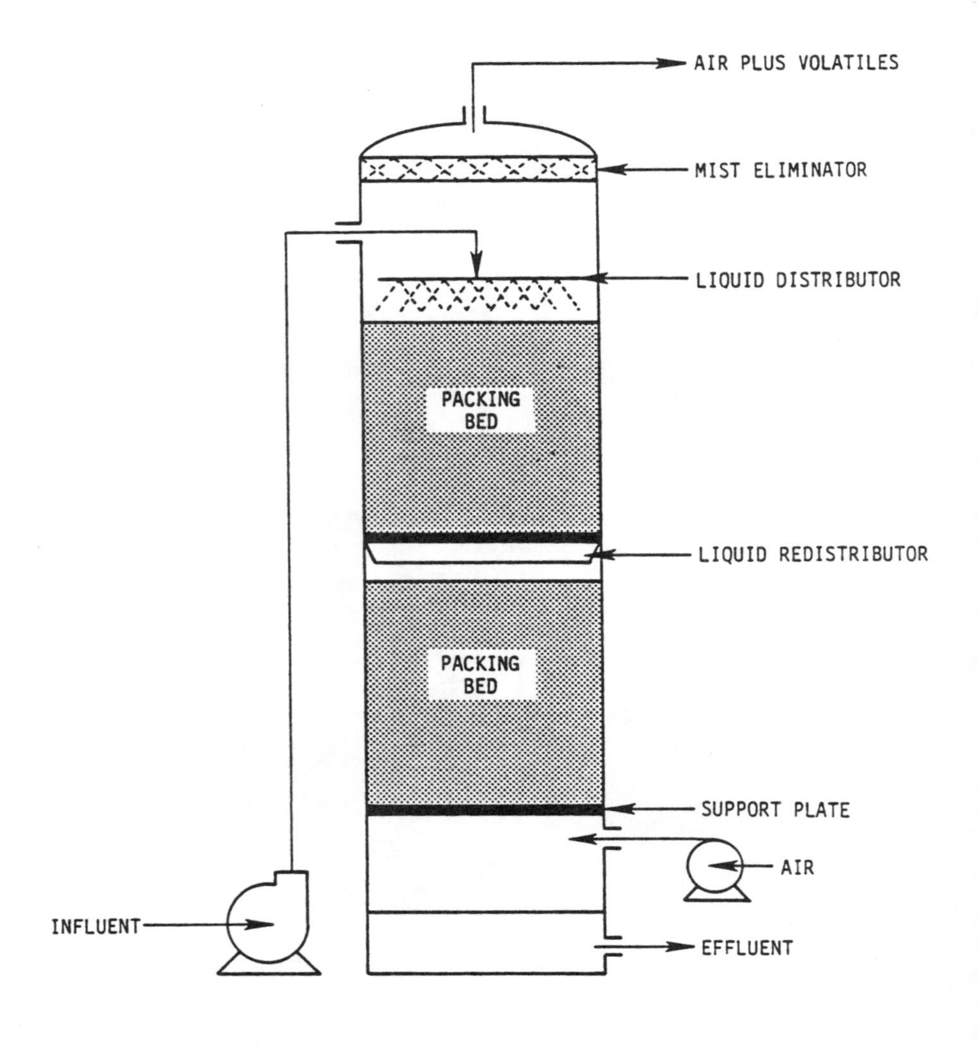

Figure 33. Schematic diagram of a countercurrent packed-tower air stripper.

Because air stripping essentially transfers volatile contaminants from the aqueous leachate to the air stream, air emission limitations for volatile organic compounds (VOC's) typically preclude discharge of this vapor directly to the atmosphere. In four of the six air stripping systems reviewed, vapor-phase carbon adsorption was incorporated to control VOC emissions. The contaminant-laden carbon must then be regenerated or disposed of, as discussed in Subsection 7.4.

Design Considerations--

Four principal factors govern the design of packed-column air-stripping towers: 1) type of packing, 2) tower diameter, 3) height of packing, and 4) air-to-water ratio.

Tower packings are available in a variety of shapes that provide high void volumes and high surface areas suitable for mass transfer. The more common packings--the Raschig ring, Berl saddle, Intalox saddle, and Pall ring--are illustrated in Figure 34. Packings may be manufactured of ceramics, metals, or plastics in nominal sizes ranging from 0.5 to 4 inches. Packings should be selected on the basis of their material characteristics (e.g., strength, durability, resistance to corrosion) and their mass-transfer characteristics (e.g., effective surface area) (Kavanaugh and Trussell 1980).

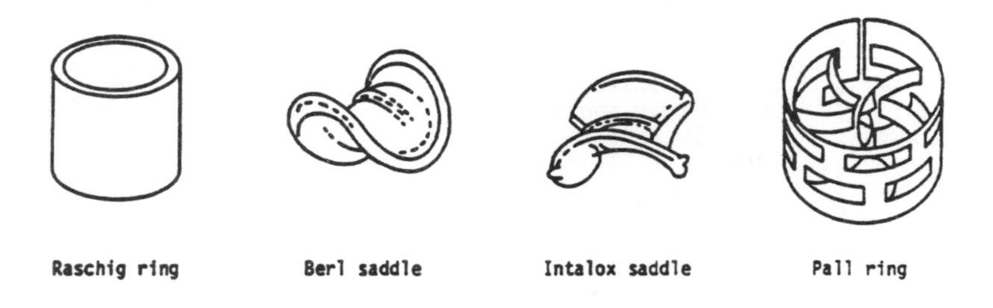

Figure 34. Common tower packings.

Flow of liquid in an air-stripping tower is influenced by tower diameter. For avoidance of poor liquid distribution, the ratio of tower diameter to nominal packing size should be no less than 8:1, and the preferred ratio is 15:1. Flow redistribution should be provided at regular intervals, with maximum acceptable distances varying from 2 to 10 times the tower diameter (Ball, Jones, and Kavanaugh 1984).

The height of packing required for various air-stripping applications depends on the desired degree of contaminant removal and the efficiency of mass transfer (Kavanaugh and Trussell 1980). The height of individual packed beds is limited by the mechanical strength of the packing and the need to redistribute the liquid (Hafslund 1979).

The volume of air required to treat a unit volume of leachate (air-to-water ratio) is a function of the volatility of the contaminants (among other factors); consequently, it must be determined through pilot-scale treatability studies. Air-to-water ratios typically range from 10 to 1 up to 300 to 1 for removal of volatile organics during warm weather operations (Metcalf & Eddy 1985). As the temperature drops, stripping becomes more difficult, and air requirements for maintaining the desired level of performance increase concomitantly (Metcalf & Eddy 1979).

Performance--

The performance of packed-tower air strippers depends on the vapor/liquid equilibrium behavior of the contaminant(s), the dimensions (height, diameter) of the tower, the efficiency of air-water contact, and the liquid temperature (Metcalf & Eddy 1979; Patterson 1985). In hazardous waste leachate applications, a minimum acceptable removal efficiency is usually defined, and a system is then designed to meet that level.

Costs--

Figure 35 presents capital and annual O&M costs of a packed-tower air stripper, and Figure 36 presents a breakdown of these costs. The capital costs of air stripping to treat flows of 25 to 100 gal/min range from $72,000 to $208,000; the packed tower makes up about 34 percent of the capital outlay. The annual O&M costs range from $38,000 to $137,000; residuals management makes up about 66 percent of these expenses. The system design assumes an air-to-water ratio of 200 to 1 (Richardson Engineering Services 1979; Barrett 1981; R. S. Means Co. 1986).

6.2.6 Steam Stripping

Process Description--

Steam stripping, or steam distillation, is a separation technique in which steam is used to remove volatile organics from leachate. The process is analogous to air stripping (Subsection 6.2.5) except that the carrier fluid is steam rather than air.

Steam stripping is carried out in a simple distillation column, as shown in Figure 37. Vapor-liquid contact may occur on discrete plates or trays (plate columns) or over continuous beds of packing (packed columns). Leachate is introduced at the top of the column, above the plates or packing, and flows downward by gravity. Steam is introduced at the bottom of the column and flows upward. Volatile components of the leachate are vaporized by contact with the rising steam and are carried overhead. The vapor (steam plus volatiles) is cooled and condensed; the condensate, which is rich in organics, is treated further or disposed of.

Applicability to Hazardous Waste Leachate--

Steam distillation is an established chemical engineering technique commonly used by industry to recover chemicals from aqueous streams or to remove contaminants from manufactured products. Although the process typically provides higher removal efficiencies than air stripping, it is considerably more expensive to operate. As a result, steam stripping is probably not practical for direct application to hazardous waste leachate treatment under normal conditions. The process may be applicable in situations where air stripping cannot effectively remove contaminants and where concentrations are too high for cost-effective carbon adsorption.

Pretreatment requirements for steam stripping are similar to those required for air stripping. Suspended solids and oil and grease should be removed to prevent fouling of the distillation equipment. Scaling can be controlled better with steam stripping than with air stripping, but corrosion rates increase rapidly with the higher temperatures, and expensive materials of construction (such as Monel steel) may be required. Operating costs are

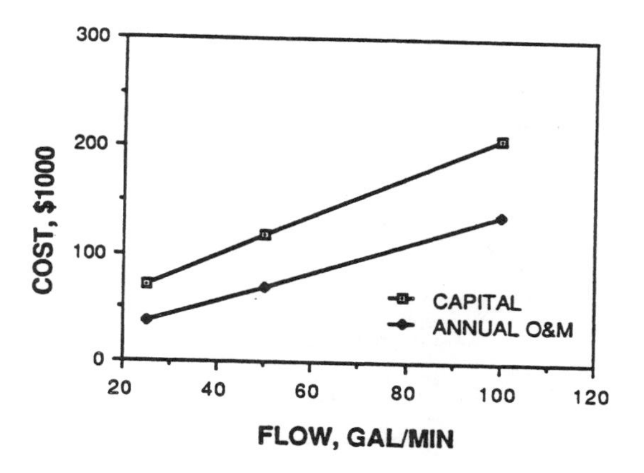

Figure 35. Capital and annual O&M costs of a packed-tower air stripper.

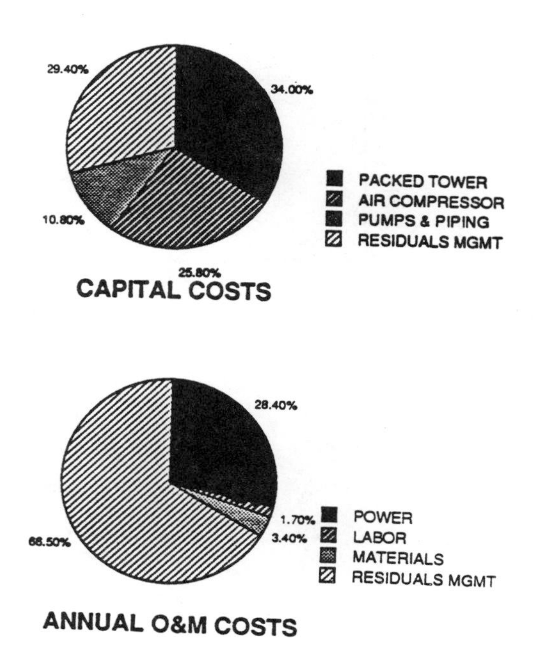

Figure 36. Breakdown of capital and annual O&M costs of a packed-tower air stripper.

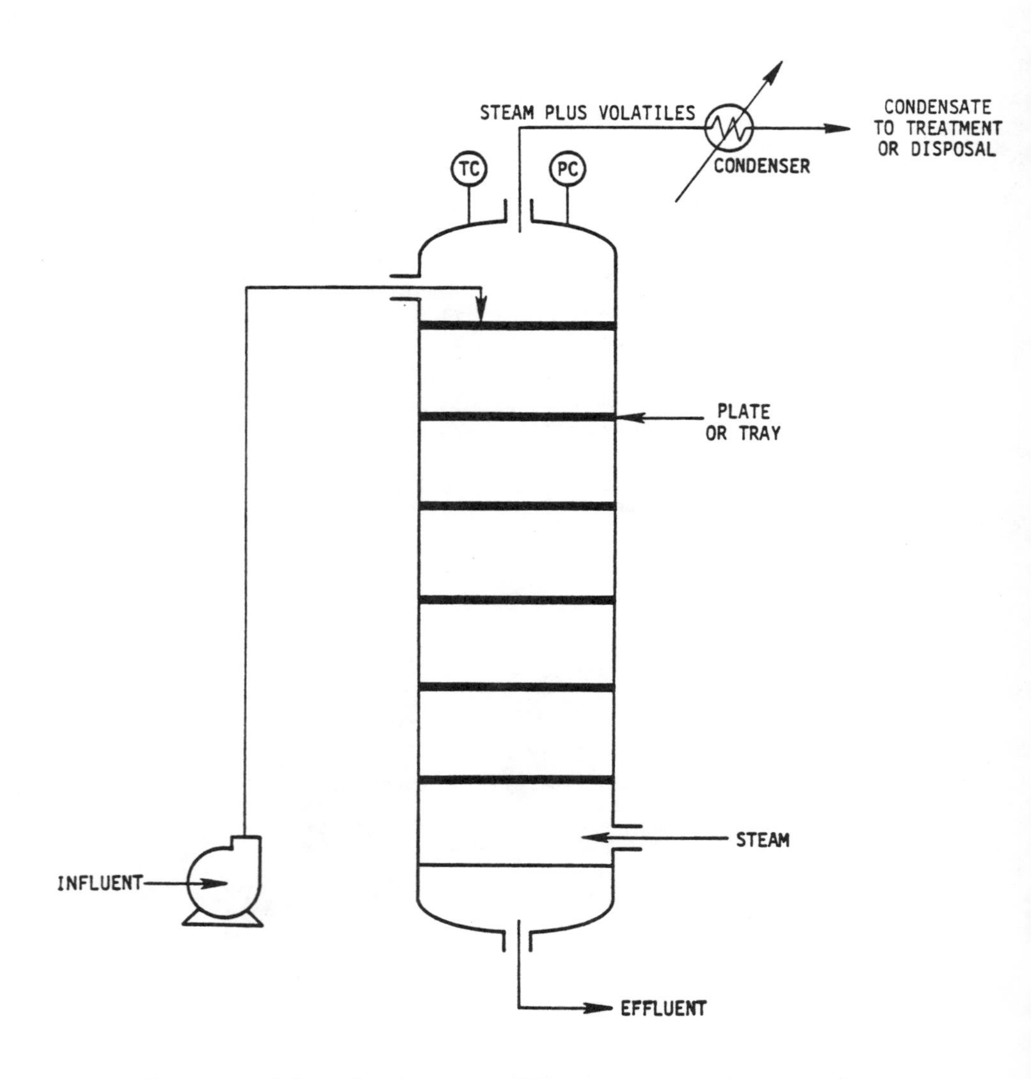

Figure 37. Schematic diagram of a plate-type steam stripping column.

also high because of the cost of steam production and the cost of treatment, incineration, or disposal of the condensate stream.

Design Considerations--

The major design considerations associated with steam stripping include selection of a plate column or packed column, the column diameter, the number of plates and spacing between plates (plate columns), the height of packing (packed columns), liquid and vapor flow rates, and operating temperature (McCabe and Smith 1976).

Packed columns are more applicable than plate columns for small-scale installations (columns less than 5 ft in diameter), severe corrosion environments, and foaming systems (Hafslund 1979). Common packing types include Raschig rings, Berl saddles, Intalox saddles, and Pall rings (see Figure 34). Common plate types include sieve plates, valve plates, and bubble-cap plates.

Performance--

Overall separation performance depends on the volatility of the constituent, the number of contacting trays or height of packing, and the ratio of the liquid flow rate to the vapor flow rate (Seader and Kurtyka 1984). Performance of a given unit can be improved by reintroduction of a portion of the recovered material (reflux) (Metcalf & Eddy 1985). At an industrial facility in Texas, steam stripping is used to reduce chlorinated hydrocarbons in leachate/contaminated ground water by 98 to 99.7 percent to less than 10 ppm.

Costs--

Figure 38 presents capital and annual O&M costs of a steam stripping column, and Figure 39 presents a breakdown of these costs. The capital costs of steam stripping for 25- to 100-gal/min streams range from $76,000 to $117,000; the packed column accounts for about 36 percent of the capital outlay. The annual O&M costs range from $37,000 to $137,000; steam accounts for about 30 percent of these expenses, and residuals management, 62 percent. The system design assumes a solvent concentration of 2 percent and the use of 0.1 lb of steam per lb of leachate (Baasel 1977; McCabe and Smith 1976; Peters and Timmerhaus 1980; Barrett 1981).

6.2.7 Reverse Osmosis

Process Description--

When two solutions of differing concentrations are separated by a semipermeable membrane (one that has a high permeability for water but a low permeability for dissolved solids), pure water from the dilute solution will pass through the membrane to the more concentrated solution until the liquid head balances the osmotic pressure (osmosis). If an external pressure greater than the osmotic pressure is applied to the concentrated solution, the direction of natural fluid flow will be reversed; i.e., pure water will flow from the concentrated solution to the dilute solution (reverse osmosis). Thus, reverse osmosis can be used to concentrate dissolved contaminants [inorganics and relatively high-molecular-weight (greater than 120) organics] in an aqueous waste stream.

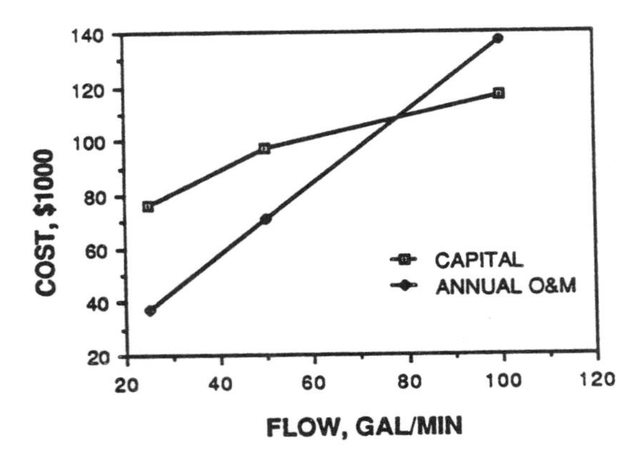

Figure 38. Capital and annual O&M costs of a steam stripping column.

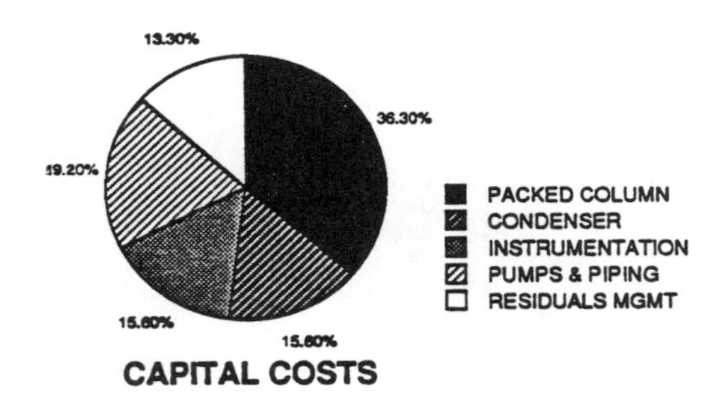

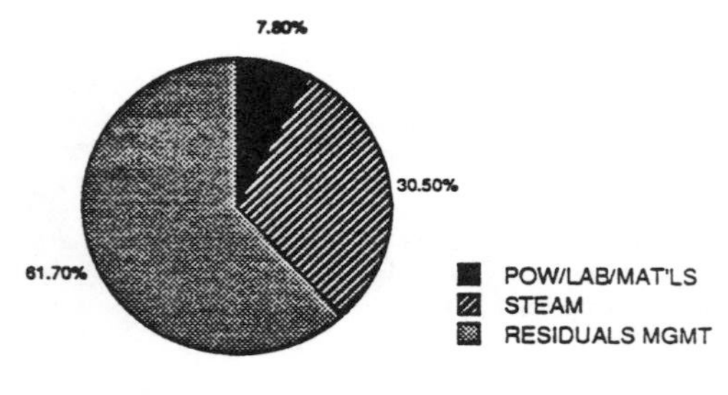

Figure 39. Breakdown of capital and annual O&M costs of a steam stripping column.

A simple schematic of a reverse-osmosis unit is shown in Figure 40. The solution or feed is pumped into a pressure vessel containing the semipermeable membrane. The purified water (permeate) is recovered at atmospheric pressure. The pressurized concentrate (reject) is let down to atmospheric pressure through a flow-regulating valve. Maximum operating pressures of reverse osmosis systems range from 400 to 1500 psig, depending on the membrane type and configuration (Patterson 1985).

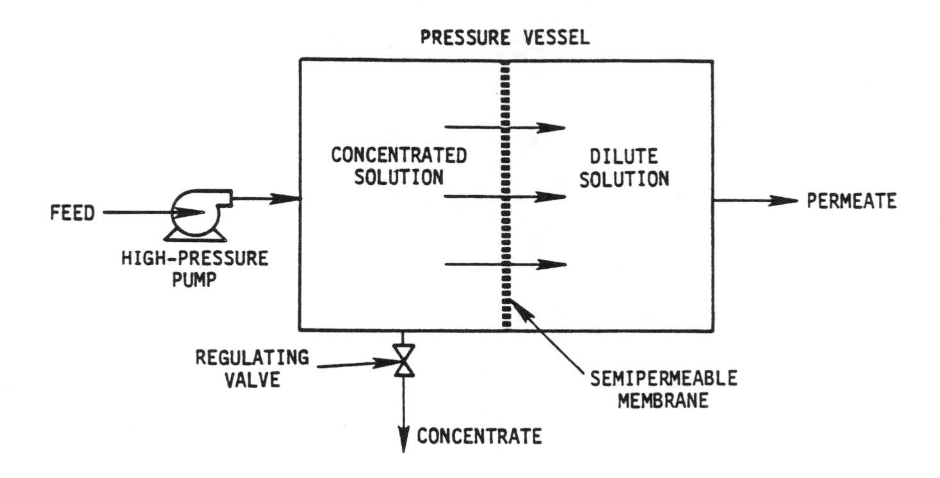

Figure 40. Schematic diagram of a reverse-osmosis unit.

All commercially available reverse osmosis membranes have an asymmetric structure--a thin, dense surface ("skin") supported by a porous substructure. This design promotes high water transport across the membrane while maintaining the ability to reject dissolved solids (Applegate 1984). Popular membrane materials include cellulose acetate, aromatic polyamides, and thin-film composites.

Because reverse-osmosis membranes are quite fragile, they have been incorporated into a variety of support devices designed to withstand the high system operating pressures. These include the tubular, spiral-wound, and hollow-fiber devices that are illustrated in Figure 41. The tubular device is a rigid-walled porous or perforated tube, 0.5 to 1 in. in diameter and lined with a membrane. The spiral-wound device consists of two to six membrane cartridges connected in series and contained within a cylindrical pressure vessel. Each cartridge is formed by winding an envelope of two sheets of

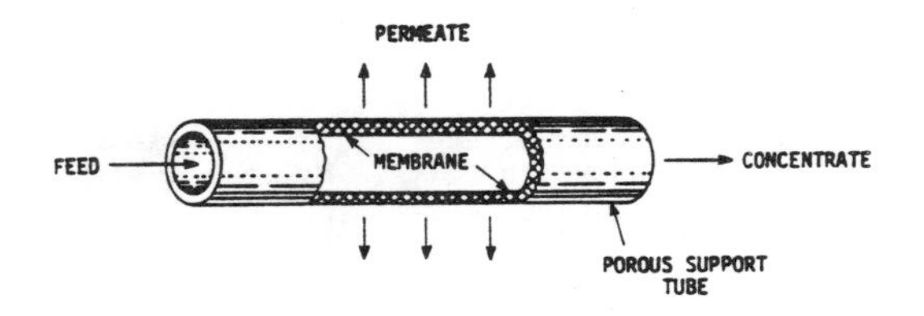

Tubular device

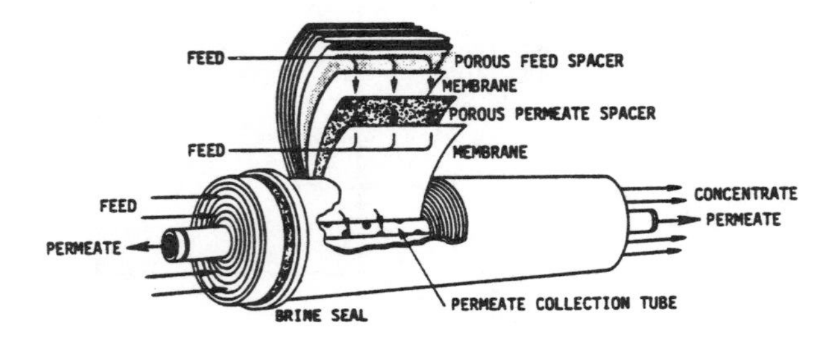

Spiral-wound device

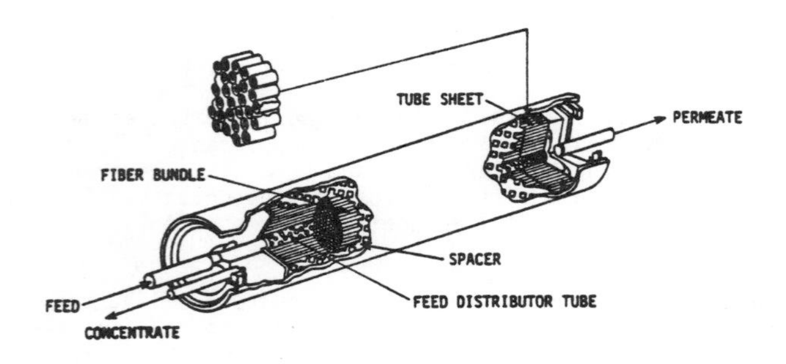

Hollow-fiber device

Figure 41. Reverse-osmosis devices.

flat membrane film, separated by porous feed and permeate spacers, around a central perforated permeate collection tube. The hollow-fiber device contains millions of fine hollow fibers (about the diameter of a human hair) externally coated with a membrane. The fiber bundle and a central feed distributor tube are encased in a cylindrical pressure vessel. Hollow-fiber and spiral-wound devices, which dominate the market, are used primarily with relatively clean feeds. The more open tubular configuration, however, may have greater applicability to complex waste streams such as leachate (Johnson 1982).

Applicability to Hazardous Waste Leachate--
Reverse osmosis is a relatively new development (within the last 25 years) in separation technology. The principal application of reverse osmosis to date has been the demineralization of brackish water and seawater to produce potable water or high-purity process water for industrial use. No evidence was found in the literature to indicate that reverse osmosis has been applied to the full-scale treatment of hazardous waste leachate, primarily because of the delicate nature of reverse-osmosis membranes and the strength and complexity of leachate. Steady progress is being made, however, in the development of durable membranes and self-cleaning reverse-osmosis units, and the potential exists for application of this technology to future hazardous waste leachate treatment systems, primarily as a polishing step subsequent to other more conventional processes (Shuckrow, Pajak, and Touhill 1982). Extensive bench- and pilot-scale studies will be needed to ascertain the feasibility of the technique.

Reverse osmosis can remove dissolved inorganics (metals, metal-cyanide complexes, and other ionic species) and high-molecular-weight organics (e.g., pesticides) from solution. Low-molecular-weight organics (e.g., phenol) will permeate the membrane as effectively as water. Reverse osmosis is generally applicable for TDS concentrations up to 50,000 mg/liter (Patterson 1985). With more highly concentrated wastes, the operating pressures required to overcome the osmotic pressure of the solution become prohibitively high. Treatment or disposal of the highly concentrated reject stream, as discussed in Subsection 7.3, will also be required.

Several factors affect the productivity and useful life of reverse-osmosis membranes:

- Compaction of the membrane due to high system pressures
- Fouling of the membrane by colloidal and suspended solids; oil, grease, and other film formers; carbonate and hydroxide precipitates; and algal and bacterial growth
- Degradation (oxidation, hydrolysis) of the membrane by chemical or biological attack
- Concentration polarization (the buildup of a boundary layer of solute at the membrane surface), which leads to precipitation of sparingly soluble salts and membrane scaling

The nature and degree of pretreatment required to protect reverse-osmosis membranes vary with the membrane material and type of device used. Normally, some combination of filtration, chlorination, carbon adsorption, and pH adjustment will be required. Specific pretreatment requirements for leachate, which may contain a number of membrane-incompatible constituents, must be determined from treatability studies. As more resistant membranes are developed, reverse osmosis will become a more technically and economically viable alternative.

Design Considerations--

The number and type of reverse-osmosis modules required for a specific application are functions of several system variables, including the desired conversion (recovery) and product flow and the required system pressure (Applegate 1984; McCoy & Associates 1985). When the desired product flow is greater than the output of a single device, multiple devices may be connected in parallel. When a greater conversion is desired, multiple devices may be connected in series. The effects of membrane compaction and fouling on productivity must be considered. Because reverse-osmosis membranes are susceptible to fouling by iron corrosion products, stainless steel and plastic materials of construction should be used.

Performance--

Removal of dissolved solutes from solution by reverse osmosis depends on the membrane material, the module configuration, and the membrane manufacturer. For example, four thin-film composite, spiral-wound membranes made by different manufacturers were tested simultaneously with chemical landfill leachate (Whittaker 1984). Major differences were observed in the performances of the different membranes, particularly with regard to their ability to remove low-molecular-weight organics and their susceptibility to fouling. The stability and long-term performance of reverse-osmosis membranes subjected to leachate remain largely unknown.

Costs--

Figure 42 presents capital and annual O&M costs of a reverse-osmosis system, and Figure 43 presents a breakdown of these costs. The capital costs of reverse osmosis for 25- to 100-gal/min streams range from $69,000 to $235,000; these costs are based on costs for a packaged system. The annual O&M costs range from $30,000 to $76,000; replacement modules account for about 8 percent of these expenses, and residuals management, 20 percent. The packaged system includes a motor pump unit, reverse-osmosis filter cartridges and housings, pressure gauges, and control modules (EPA 1982a; personal communication from B. Thomas, Continental Water Co., Dallas, Texas, June 9, 1986).

6.2.8 Ultrafiltration

Process Description--

Ultrafiltration is a membrane process capable of separating solution components on the basis of molecular size, shape, and flexibility. The semipermeable membrane acts as a sieve to retain dissolved and suspended macromolecules that are physically too large to pass through its pores, which range in size from 0.001 to 0.02 μm (Applegate 1984). As in reverse osmosis

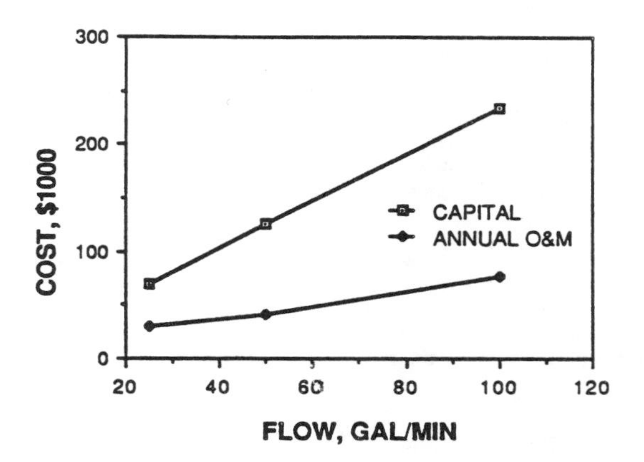

Figure 42. Capital and annual O&M costs of a reverse-osmosis system.

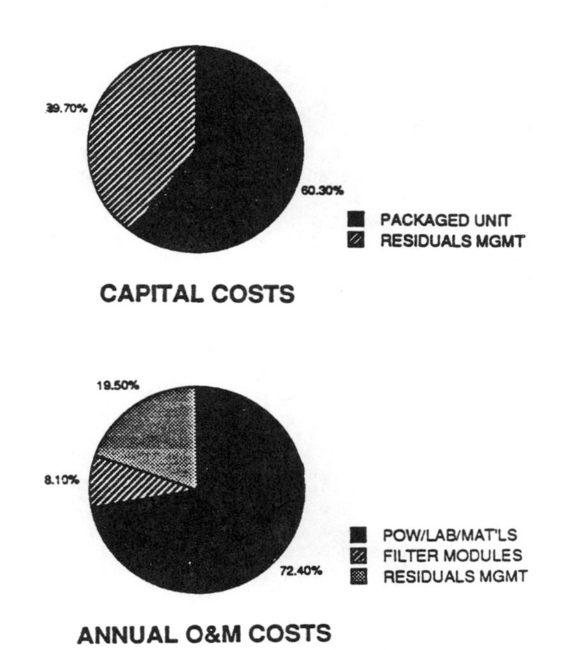

Figure 43. Breakdown of capital and annual O&M costs of a reverse-osmosis system.

(described in Subsection 6.2.7), pressure is the driving force for the separation. The osmotic pressures associated with solutions of macromolecules, however, are much lower than those associated with salt solutions; thus, ultrafiltration systems operate at significantly lower pressures (10 to 100 psig) than reverse-osmosis systems do (Applegate 1984). A simple schematic of an ultrafiltration system would resemble that of the reverse-osmosis process (Figure 40), except that a low-pressure pump would be used in place of the high-pressure pump.

Ultrafiltration membranes have an asymmetric structure--a thin, dense skin supported by a porous substructure--which promotes high productivity. Membranes can be made from a variety of polymers, the most common of which are cellulose acetate and polysulfone. In general, ultrafiltration membranes have a more open structure than do reverse osmosis membranes. The retention capability of an ultrafiltration membrane is often described in terms of its "molecular weight cutoff," i.e., the molecular weight of the smallest solute that is retained by the membrane. Commercially available membranes have molecular weight cutoffs ranging from 500 to 1,000,000 (McCoy & Associates 1985).

The primary ultrafiltration devices are the tubular, spiral-wound, and hollow-fiber devices. The tubular and spiral-wound configurations are similar to reverse-osmosis devices in design and operation (Figure 41). Hollow fibers used in ultrafiltration, however, have the membrane on the inside and are significantly larger in diameter than those used in reverse osmosis. Feed flows through the fiber bores, and permeate is collected on the outside.

Applicability to Hazardous Waste Leachate--

Ultrafiltration has several industrial applications, but this process (like reverse osmosis) has not yet been applied to the full-scale treatment of hazardous waste leachate. Extensive bench and pilot studies would be required prior to any such application to assure long-term stability and durability of the membrane. It has been suggested that ultrafiltration may be effective as a pretreatment process for reverse osmosis in the removal of large organics, including PCB's and pesticides, from aqueous leachate streams. The greatest potential for application of these membrane technologies probably involves sites where leachate contains only one primary contaminant. As membranes exhibiting greater productivity and chemical resistance are developed, ultrafiltration will likely become a more viable treatment alternative.

Ultrafiltration generally removes high-molecular-weight (greater than 500) species from solution, including macromolecules (proteins, polymers), complexed metals, oil emulsions, colloidal dispersions (clay, microorganisms), and suspended solids; low-molecular-weight species (organics, inorganic salts) will not be retained by the membrane (Henry et al. 1984; McCoy & Associates 1985).

Concentration polarization (the buildup of rejected solutes at the membrane surface) is more of a problem in ultrafiltration systems than in reverse-osmosis systems. The effects of concentration polarization (decreased productivity) can be minimized by maintaining a high-velocity fluid flow across the membrane, which tends to shear off some of the polarization gel layer.

Minimum pretreatment requirements for ultrafiltration include equalization, oil/water separation, and suspended solids removal through sedimentation and/or filtration. Additional pretreatment requirements, as determined through treatability studies, may include pH adjustment or removal of contaminants that could foul or damage the membrane. Treatment and/or disposal of the concentrate (as discussed in Subsection 7.3), is also required.

Design Considerations--

The number and type of ultrafiltration modules required for a specific application are functions of several system variables, including the desired conversion (recovery), separation, and product flow (Applegate 1984). Maintenance of high productivity (minimized concentration polarization) with ultrafiltration membranes requires high-velocity fluid flow. At such velocities, however, conversion may be as low as 10 percent. Therefore, recirculation of the feed stream is required. Recirculation may be achieved with either batch or continuous operation, as illustrated schematically in Figure 44. With batch operation, the feed is recirculated to a feed tank, and the batch is processed until the desired volume reduction (conversion) is obtained. With continuous operation, feed is added to and concentrate is bled from a continuous recirculation loop at equal rates. The concentrate from this stage can then be fed to subsequent stages as needed to effect the desired degree of separation.

Performance--

Treatment of industrial landfill leachate by use of ultrafiltration has been investigated at the laboratory scale. Slater (1982) found that 80 to 85 percent of the organic matter present permeated an ultrafiltration membrane with a molecular weight cutoff of 500. Rapid membrane fouling by residual suspended solids and concomitant reduction in productivity were observed by Syzdek and Ahlert (1984) for membranes with molecular weight cutoffs of 50,000 and 300,000. This fouling layer could be minimized by increasing the shear at the membrane surface or by lowering the strength of the leachate.

Costs--

Figure 45 presents capital and annual O&M costs of an ultrafiltration system, and Figure 46 presents a breakdown of these costs. Capital costs for ultrafiltration for 25- to 100-gal/min streams range from $55,000 to $134,000; these costs are based on costs for a packaged system. The annual O&M costs range from $32,000 to $93,000; replacement modules account for about 31 percent of these expenses, and residuals management, 6 percent (EPA 1982a; personal communication from B. Thomas, Continental Water Co., Dallas, Texas, June 9, 1986).

6.2.9 Ion Exchange

Process Description--

Ion exchange is a process that reversibly exchanges ions in solution with ions of like charge retained on an insoluble resinous solid called an ion-exchange resin. The ion-exchange resin has the ability to exchange either positively charged ions (cation exchange) or negatively charged ions (anion exchange). The exchangeable ions (H^+ or Na^+ in the case of cation exchange, OH^- or Cl^- in the case of anion exchange) are held to the resin, a

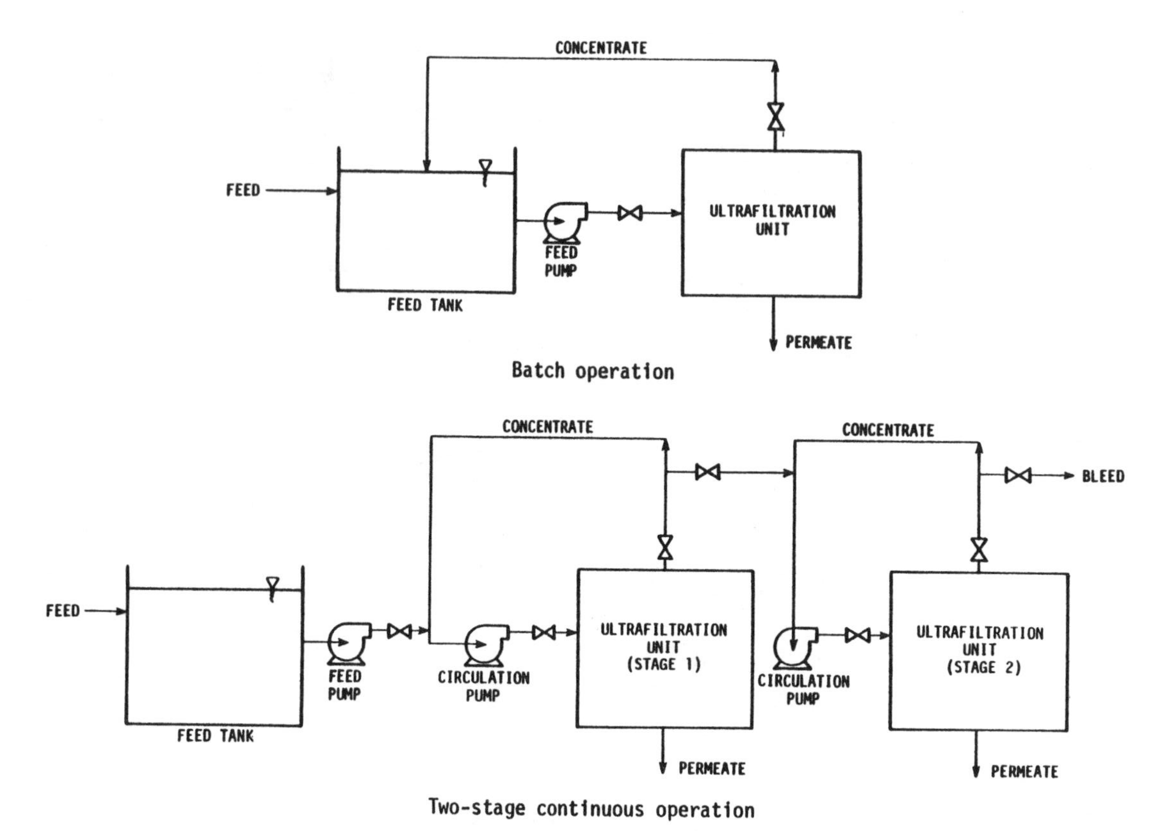

Figure 44. Batch and continuous ultrafiltration operations.

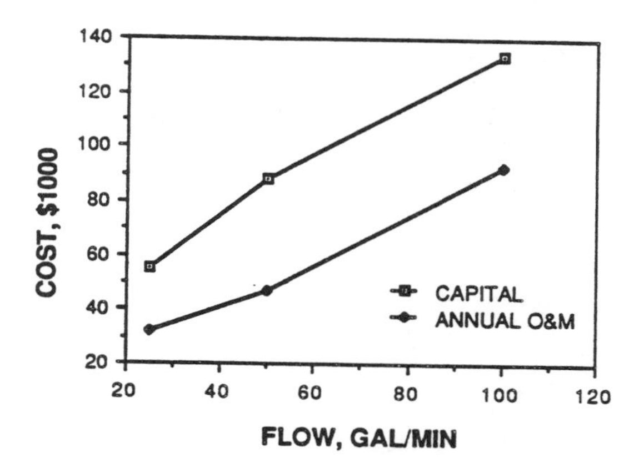

Figure 45. Capital and annual O&M costs of an ultrafiltration system.

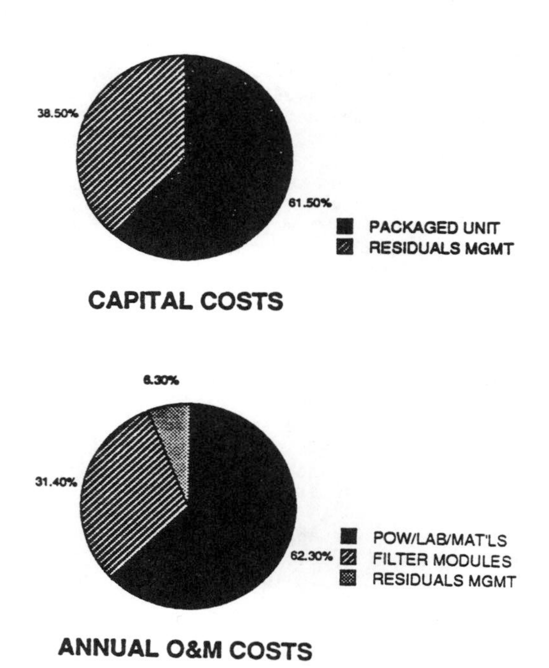

Figure 46. Breakdown of capital annual O&M costs of an ultrafiltration system.

synthetic organic polymer or natural zeolite containing fixed ions of opposite charge, by relatively weak electrostatic forces and can be readily displaced by ions in leachate with a greater affinity for the resin. The ion-exchange process may be represented by the following equilibrium equation:

$$Na_2R + Ca^{+2} \rightleftarrows CaR + 2Na^+$$

where calcium ions in solution displace sodium ions initially on the resin (R).

When the useful exchange capacity of the resin is exhausted, the resin may be regenerated by washing it with a solution containing an excess of the ion initially adsorbed on the solid (i.e., a strong acid for hydrogen-form cation exchangers, a strong base for hydroxyl-form anion exchangers, and a concentrated salt solution for sodium- and chloride-form exchangers). Regeneration essentially reverses the exchange process, and ions initially present in the leachate are concentrated in the regenerant solution. The regeneration process may be represented by the following equation:

$$CaR + 2NaCl \rightleftarrows Na_2R + CaCl_2$$

where a sodium chloride solution is used to regenerate a sodium-form cation exchange resin.

Ion exchange (cationic and anionic) may be carried out in separate fixed-bed columns operated in series or, less commonly, in a single vessel (mixed-bed column or continuous countercurrent column) that contains both resin types. Fixed-bed columns (illustrated in Figure 47) are typically operated in a downflow mode. Leachate is introduced at the top of the column under pressure, flows downward through the resin bed, and is discharged at the bottom. When the exchange capacity of the resin is exhausted, the bed is first backwashed to remove trapped solids and then regenerated with an appropriate chemical solution. The bed may be regenerated cocurrently or countercurrently to the service flow; however, the latter method is generally more effective. Finally, the resin is rinsed to remove excess regenerant prior to initiation of the next work cycle.

Applicability to Hazardous Waste Leachate--

Ion exchange is used primarily for the removal of dissolved ionic species when a high-quality effluent is required. The technology is widely used for domestic-water softening, boiler-water deionization, and treatment of metal electroplating wastes. The applicability of this process to the treatment of hazardous waste leachate is probably limited to use as a final polishing stage where effluent is discharged to sensitive surface waters. No evidence has been found that ion exchange has been applied to the full-scale treatment of hazardous waste leachate.

Technically, ion exchange is applicable to the removal of a broad range of inorganics, including metals, halides, and cyanides (EPA 1982a; Patterson 1985). Ion exchange is not suitable for removal of high concentrations of

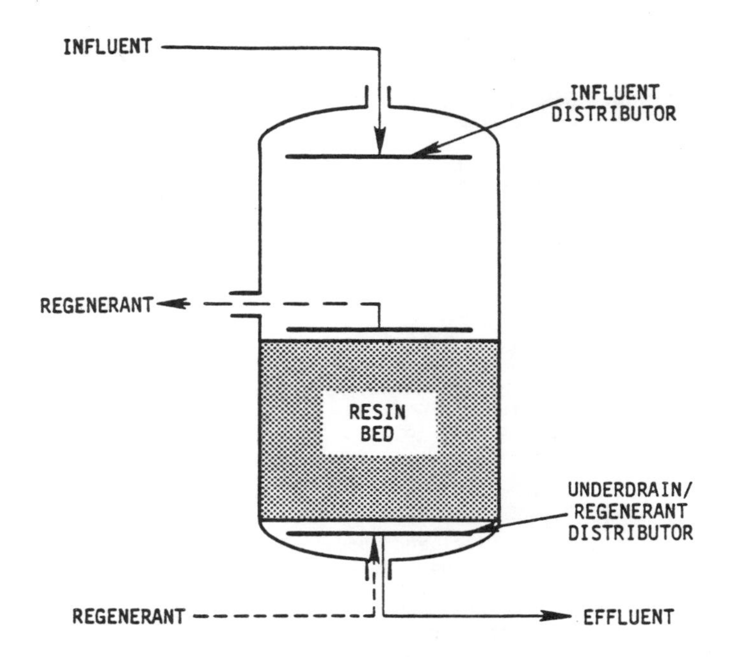

Figure 47. Schematic diagram of a downflow, fixed-bed, ion exchanger with countercurrent regeneration.

dissolved solids because the exchange resin is rapidly exhausted, and costs for regeneration become prohibitively high. A practical upper concentration limit for exchangeable ions for efficient operations is about 2500 mg/liter as $CaCO_3$ or 0.05 equivalents/liter (EPA 1982a).

Extensive pretreatment is required prior to ion exchange in a hazardous waste leachate treatment system. Suspended solids and nonaqueous liquids (oil, grease, etc.) must be removed to prevent fouling and plugging of the resin beds. High concentrations of ionic species should be removed through the less costly process of precipitation/flocculation/sedimentation. Strong oxidizing agents (including excess chlorine) and some organic compounds (particularly aromatics), which can irreversibly bind to the resin, must also be removed prior to ion exchange.

The periodic regeneration of ion-exchange resins results in a contaminant-laden waste stream that requires further treatment or disposal. Alternatives for handling this concentrated liquid waste are addressed in Subsection 7.3.

Design Considerations--

Selection of an appropriate resin is a primary consideration in the design of an ion-exchange system. Exchange resins exhibit an order of selectivity or affinity for ions. Ions with a high selectivity are preferentially adsorbed over ions with a low selectivity.

The usable exchange capacity of a given resin varies under differing operating conditions; therefore, it must be determined experimentally in laboratory or pilot-plant studies. Optimum regenerant quantities must also be determined experimentally. These parameters, together with the characteristics of the leachate to be treated and the required flow rate, determine the volume of resin needed (Miller et al. 1984).

Performance--

No data are available regarding the reliability or effectiveness of ion exchange in hazardous waste leachate treatment applications. Total dissolved solids removal efficiencies of 90 to 99 percent have been reported for treatment of water supplies with a conventional two-stage (anionic and cationic) exchange system (Metcalf & Eddy 1979). Comparable efficiencies have been reported for removal of metals from various industrial wastewaters (Patterson 1985).

Costs--

Figure 48 presents capital and annual O&M costs of an ion-exchange unit, and Figure 49 presents a breakdown of these costs. The capital costs of ion exchange for 25- to 100-gal/min streams range from $59,000 to $118,000; the equipment makes up about 27 percent of the capital outlay, and the resin, 13 percent. The annual O&M costs range from $10,000 to $26,000; makeup resin accounts for about 13 percent of these expenses and residuals management, 50 percent. The system includes two pressure vessels with internals, the exchange resins (anion and cation), pumps and piping, regenerating chemicals, and storage facilities (EPA 1982a; Heister 1973; Chemical Marketing Reporter 1986).

6.2.10 Wet-Air Oxidation

Process Description--

Wet-air oxidation (WAO) is the aqueous-phase oxidation of concentrated organic and inorganic wastes in the presence of oxygen at elevated temperature and pressure. Pressure in the range of 2,170 to 20,710 kPa (300 to 3000 lb/in²) is required to maintain water in its liquid state, which allows oxidation to progress at lower temperatures than would be required for open-flame combustion. Water serves to moderate the oxidation rate by absorbing excess heat of reaction. Reactor temperatures typically range from 175° to 320°C (350° to 610°F) (Dietrich, Randall, and Canney 1985).

In the WAO process (Figure 50), the influent is pumped under high pressure through a series of heat exchangers to preheat the feed. The source of heat to the exchangers is provided by hot oxidized effluent or an auxiliary heat supply (e.g., a hot-oil heater). Preheated influent enters the pressurized reactor with compressed air or high-pressure pure oxygen and reacts for a period ranging from a few minutes to several hours. From the reactor,

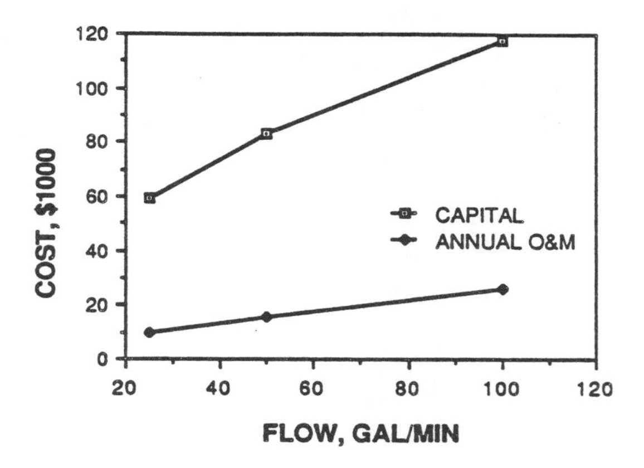

Figure 48. Capital and annual O&M costs of an ion-exchange unit.

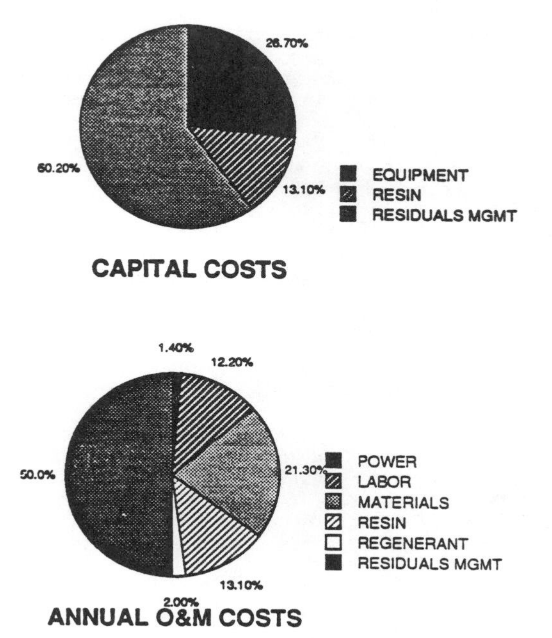

Figure 49. Breakdown of capital and annual O&M costs of an ion-exchange unit.

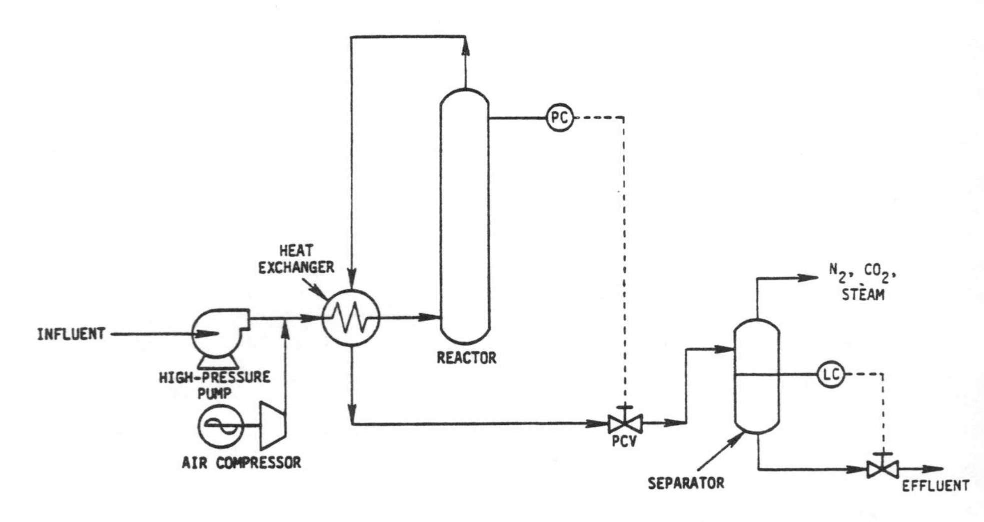

Figure 50. Schematic flow diagram of the wet-air oxidation process.

the hot oxidized effluent is cooled by heat exchange with the feed before exiting through a pressure-reducing station. After pressure letdown, the vapor and liquid components of the cooled effluent are separated. The vapor passes through an air pollution control device and is vented to the atmosphere. The liquid is discharged to a subsequent treatment process.

Applicability to Hazardous Waste Leachate--

The wet-air oxidation process may be applied to any concentrated organic or inorganic waste stream with a COD between 10,000 and 100,000 mg/liter (Metcalf & Eddy 1985). It is particularly suitable for waste streams that are too dilute for incineration but too refractory for chemical or biological oxidation. Wet-air oxidation has been applied on a bench- or pilot-scale basis to numerous industrial wastewaters and sludges containing hazardous organic constituents (Dietrich, Randall, and Canney 1985). The areas of greatest potential applicability for hazardous waste leachate appear to be treatment of concentrated liquid waste streams generated by steam stripping, ultrafiltration, or reverse osmosis; treatment of biological waste sludges; and regeneration of powdered activated carbon (Shuckrow, Pajak, and Touhill 1982).

Through a series of oxidation and hydrolysis reactions, wet-air oxidation alters hazardous leachate constituents to partially oxidized intermediates or to innocuous end products if the reaction is taken to completion (EPA 1982a). Reduced-sulfur compounds, such as sulfides and mercaptans, are oxidized to inorganic sulfates; cyanides, to carbon dioxide and ammonia or nitrogen gas; chlorinated hydrocarbons, to hydrochloric and simple organic acids; and metals, to metal salts. Certain highly chlorinated aromatic organics, such as PCB's, are resistant to destruction (Metcalf & Eddy 1985).

Oxidized effluent from a WAO unit generally requires further treatment (biological or physical/chemical) to meet effluent COD discharge limitations. Hydrochloric acid, which is generated by the oxidation of chlorinated hydrocarbons, must be scrubbed from the off-gas. Oxidized sulfur and nitrogen species are retained in the liquid phase; thus, SO_x and NO_x do not pose air pollution problems (Dietrich, Randall, and Canney 1985). Because of the complexity of the process, skilled labor is required to operate a WAO unit.

Design Considerations--

The degree of oxidation of organics and inorganics in a WAO unit is primarily a function of reaction temperature and residence time. Oxidation is accompanied by temperature rise, and the heat produced can be used to sustain the process. For oxidation to proceed autogenously (i.e., without auxiliary fuel), the COD of the feed must be greater than about 15,000 mg/liter (Dietrich, Randall, and Canney 1985). Additional heat can be supplied as needed by injecting steam directly into the reactor. Stainless steel or other corrosion-resistant materials of construction should be used for protection against the corrosive atmosphere that results from the oxidation of chlorinated hydrocarbons. Scale buildup in the heat exchangers can be removed by periodical flushing of the system with a solution of nitric acid. Skid-mounted package WAO units are available for flows of 15,000 gal/day or less.

Performance--

Table 6 presents the results of extensive bench-scale studies of the wet-air oxidation of organic priority pollutants (pure compounds). In most cases, results indicate greater than 99 percent destruction of the individual compounds (Dietrich, Randall, and Canney 1985). General trends observed in these studies (i.e., good removal efficiencies for chlorinated aliphatic hydrocarbons; moderate removal efficiencies for chlorinated aromatic hydrocarbons; better removal efficiencies with higher temperatures, longer residence times, and catalyzed reactions) were also observed in studies with waste mixtures. No performance data are available on the wet-air oxidation of hazardous waste leachate.

Costs--

Figure 51 presents capital and annual O&M costs of a wet-air oxidation unit, and Figure 52 presents a breakdown of these costs. The capital costs of wet-air oxidation for 25- to 100-gal/min streams range from $421,000 to $826,000; the reactor vessel accounts for about 47 percent of the capital outlay. The annual O&M costs range from $82,000 to $269,000; power makes up about 77 percent of these expenses. The system was designed for a 1-h retention time. The reactor is rated at 1500 lb/in² (Peters and Timmerhaus 1980; Guthrie 1969; Richardson Engineering Services 1979; Zimpro 1984).

TABLE 6. BENCH-SCALE WET-AIR OXIDATION OF PRIORITY POLLUTANTS (PURE COMPOUNDS)[a]

Compound	Operating conditions, °C/min	Starting concentration, mg/liter	Final concentration, mg/liter	Destruction efficiency, %
Acenaphthene	275/60	7,000	0.5	99.99
Acrolein	275/60	8,410	<3	>99.96
Acrylonitrile	275/60	8,060	80	99.0
2-Chlorophenol	275/60[b]	12,410	15	99.88
2,4-Dimethylphenol	275/60	8,220	0.1	99.99
2,4-Dinitrotoluene	275/60	10,000	26	99.74
1,2-Diphenylhydrazine	275/60	5,000	6	99.88
4-Nitrophenol	275/60	10,000	40	99.6
Pentachlorophenol	275/60[b]	5,000	135	97.3
Phenol	275/60	10,000	20	99.8
Formic acid	300/60	25,000	410	98.3
Chloroform	275/60	4,450	3	99.9
Carbon tetrachloride	275/60	4,330	12	99.7
1,2-Dichloroethane	275/60	6,280	13	99.8
N-Nitrosodimethylamine	275/60	5,030	22	99.6
Hexachlorocyclopentadiene	300/60	10,000	<15	>99.9
Toluene	275/60	4,330	12	99.7
Nitrobenzene	320/120[b]	5,125	255	95.0
Chlorobenzene	275/60[b]	5,535	1550	72.0
Pyridine	320/120[b]	3,910	570	85.4
2,4-Dichloroaniline	275/60[b]	259	<0.5	>99.8
2,4,6-Trichloroaniline	320/120	10,000	2.5	99.97
Dibutylphthalate	275/60	5,230	26	99.5
Isophorone	275/60	4,650	29	99.4
1-Chloronaphthalene	275/60[b]	5,970	5	99.92
Pyrene	275/60	500	0.26	99.95
Malathion	250/60	11,800	18	99.85
Kepone	280/60[b]	1,000	690	31.0
Arochlor 1254	320/120	20,000	7400	63.0
1,2-Dichlorobenzene	320/60[b]	6,530	2017	69.1

[a] Source: Dietrich, Randall, and Canney 1985.

[b] Catalyzed.

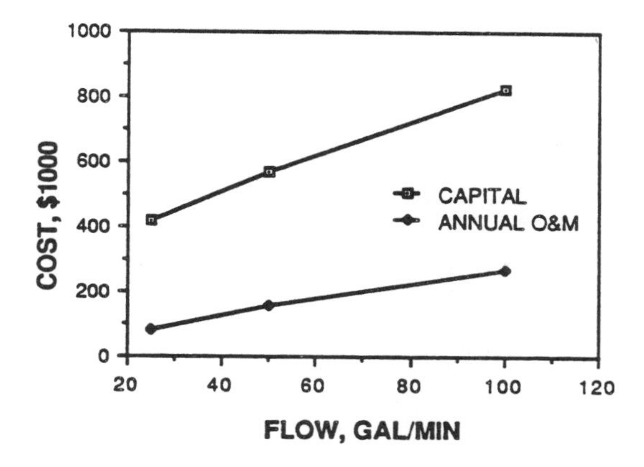

Figure 51. Capital and annual O&M costs of a wet-air oxidation unit.

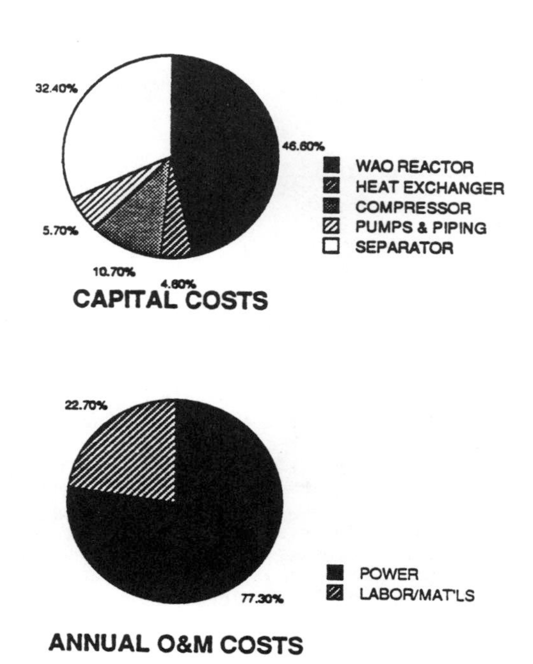

Figure 52. Breakdown of capital and annual O&M costs of a wet-air oxidation unit.

6.3 BIOLOGICAL TREATMENT OPERATIONS

6.3.1 Activated Sludge

Process Description--

The activated-sludge process is a suspended-growth, biological treatment process that uses aerobic microorganisms to biodegrade organic contaminants in leachate. With conventional activated-sludge treatment, the leachate is aerated in an open tank with diffusers or mechanical aerators. This provides the microorganisms with oxygen to oxidize biodegradable compounds present in the waste to NO_3, SO_4, CO_2 and H_2O (EPA 1980). After the aeration phase, the mixed liquor (the mixture of microorganisms and leachate) is pumped to a gravity clarifier to settle out the microorganisms. A high percentage of the settled biomass is recycled to the aeration tank to maintain the design mixed-liquor suspended solids level, and the excess sludge is wasted. A schematic flow diagram of the conventional activated-sludge process is presented in Figure 53.

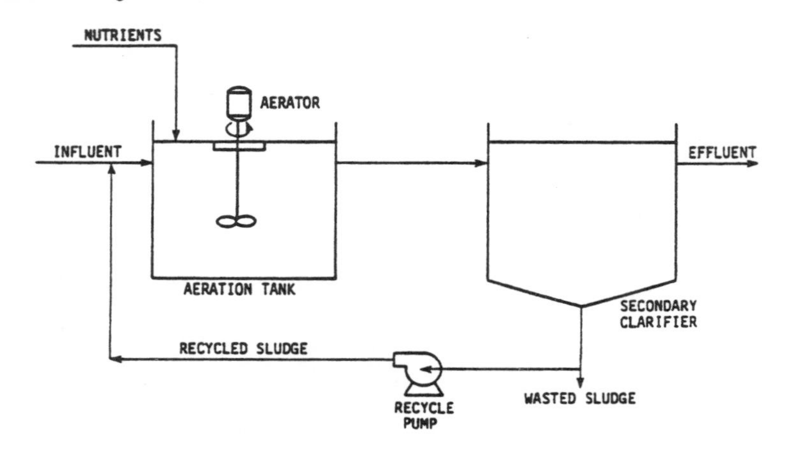

Figure 53. Schematic flow diagram of the conventional activated-sludge process.

Variations in the conventional activated-sludge process have been developed to provide greater tolerance for shock loadings, to improve sludge settling characteristics, and to achieve higher BOD_5 removals. Process modifications include complete mixing, step aeration, modified aeration, extended aeration, contact stabilization, and the use of pure oxygen (EPA 1982a).

Applicability to Hazardous Waste Leachate--

In general, biological processes are the most cost-effective means for reducing the organic content of leachate, particularly when complete onsite treatment is required. Four of the six leachate-treatment case-study sites presented in Table 4 (p. 21) that specify ground- or surface-water discharge of treated leachate also specify onsite biological treatment, and three of these four specify the activated-sludge method.

A practical upper limit for influent BOD_5 to an activated-sludge system is 10,000 mg/liter. Because of the sensitivity of the system, activated sludge should be preceded by equalization to buffer hydraulic and organic load variations; precipitation/flocculation/sedimentation to remove metals and suspended solids; neutralization to adjust the pH to near neutral; and nutrient addition to provide adequate levels of nitrogen, phosphorus, and trace elements. Leachate containing a high fraction of nonbiodegradable (refractory) organics will require post-treatment by carbon adsorption prior to final effluent discharge. Filtration may also be provided to remove any residual suspended solids.

Process residuals from activated-sludge treatment of leachate include waste activated sludge, which may contain high concentrations of refractory organics removed by adsorption onto solids, and air emissions of VOC's that are stripped from the waste during aeration. Management alternatives for sludge and VOC air emissions are addressed in Section 7.

Design Considerations--

Key design parameters for an activated-sludge system include the organic loading (lb BOD_5/day per 1000 ft^3), the hydraulic retention time (HRT, h) the mixed-liquor suspended solids concentration (MLSS, mg/liter), the food-to-microorganism ratio (F:M, lb BOD_5/day per lb MLVSS), the air requirement (ft^3/min per lb BOD_5 removed), and the solids residence time (SRT, d). Limitations of the conventional activated-sludge process (poor tolerance for shock loads, low acceptable organic loadings) can result in poor system performance for leachate treatment applications. These limitations can be overcome through the use of one of several process modifications. The completely mixed activated-sludge process, in which influent waste and return sludge are introduced at several points in the aeration tank, reduces the effect of shock loads of organics. The extended-aeration process, which involves long detention times and a low F:M ratio, provides a high degree of oxidation and a minimum of excess sludge. Pure-oxygen processes operate at higher organic loadings and lower aeration requirements (EPA 1982a). For successful winter operation of any activated-sludge system, a covered tank and a supplemental heat source are required.

Performance--

In general, the activated-sludge process can readily degrade simple organic species such as alkanes, alkenes, and aromatics. Halogenated hydrocarbons are degraded more slowly (Blaney 1986). Some halogenated organics that have been successfully treated by the activated-sludge process include chlorobenzene, carbon tetrachloride, chlorodibromoethane, methyl chloride, and 2,4,6-trichlorophenol (Turner 1986). The performance of an activated-sludge system is related to the degree of acclimation of the biomass. The use of indigenous bacteria from the waste-disposal site can speed reaction rates and improve total system performance.

Costs--

Figure 54 presents capital and annual O&M costs of a conventional activated-sludge system, and Figure 55 presents a breakdown of these costs. The capital costs of activated sludge to treat flows of 25 to 100 gal/min range from $184,000 to $364,000; the aeration basin accounts for about 28 percent of the capital outlay, and the clarifier, 30 percent. The annual O&M costs range from $18,000 to $47,000; residuals management makes up about 60 percent of these expenses. The aeration basin design assumes a detention time of 6 h and an aerator power input of 0.1 hp per 1000 gal. The clarifier design is based on an operation of 600 gal/day per ft^2 (Peters and Timmerhaus 1980; Guthrie 1969; Richardson Engineering Services 1979; Barrett 1981; U.S. Army Corps of Engineers 1985; EPA 1982a).

6.3.2 Sequencing Batch Reactor

Process Description--

The sequencing batch reactor (SBR), shown in Figure 56, is a fill-and-draw activated-sludge system. Unlike conventional, continuous-flow, activated-sludge systems, which have separate tanks for equalization, aeration, and clarification, the SBR performs all operations in a single tank. Each cycle of the batch operation involves five phases of treatment in timed sequence, as illustrated in Figure 57 and described below:

Fill. Leachate is fed to the SBR, which contains an acclimated biomass from the previous cycle. Aeration may or may not be provided during the fill phase.

React. The reactor contents are actively mixed and aerated to allow the microorganisms to aerobically degrade the organic matter present in the leachate.

Settle. Mixing and aeration are stopped, and the suspended solids are allowed to settle under quiescent conditions.

Draw. Clarified supernatant is withdrawn from the reactor for further treatment and discharge.

Idle. Settled solids are retained in the reactor for the next cycle. A portion of the settled sludge may be wasted during the idle phase.

Applicability to Hazardous Waste Leachate--

The sequencing batch reactor, like the conventional activated-sludge process described in Subsection 6.3.1, can be used to biodegrade organic contaminants (e.g., phenol) in leachate. The SBR is particularly applicable to the treatment of leachate that is not generated in sufficient volume to justify a continuous-flow process. With an SBR, the leachate can be accumulated in a holding tank for intermittent treatment. The SBR also has greater operational flexibility to accommodate changing feed characteristics (flow and/or organic loading) and can achieve more complete treatment through adjustment of reaction parameters than the conventional activated-sludge system (McCoy & Associates 1986; Ying et al. 1986). The SBR is currently used to treat hazardous waste leachate at Occidental Chemical Corporation's Hyde Park

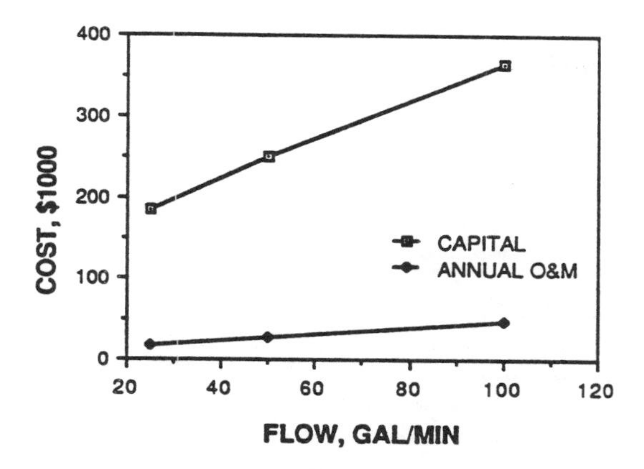

Figure 54. Capital and annual O&M costs of a conventional activated-sludge system.

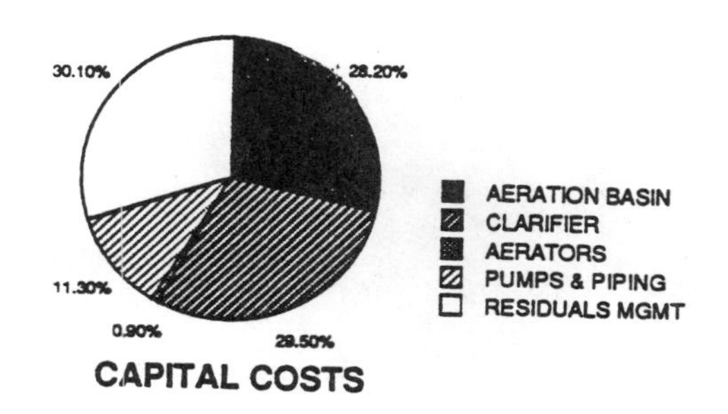

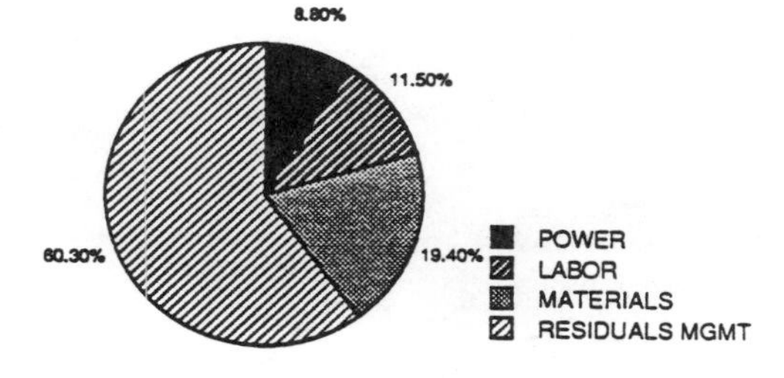

Figure 55. Breakdown of capital and annual O&M costs of a conventional activated-sludge system.

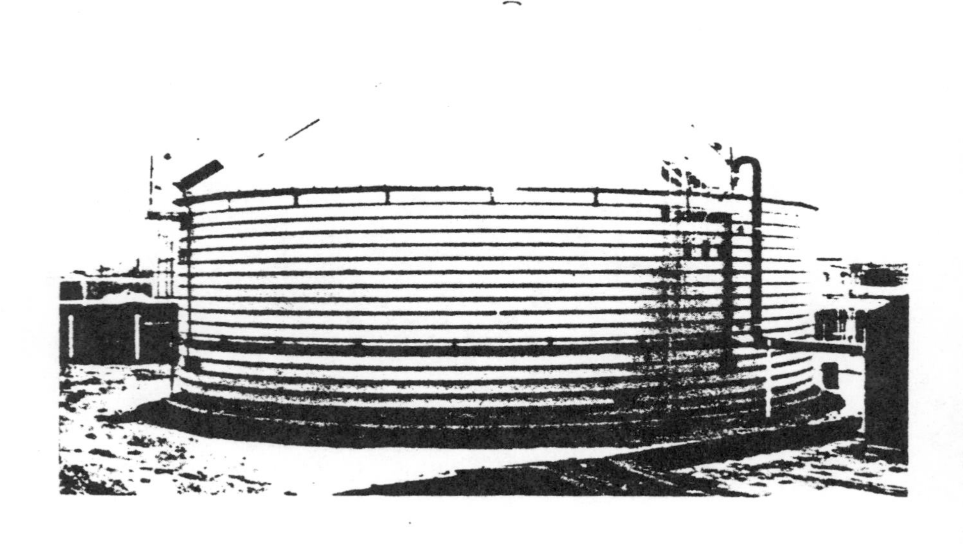

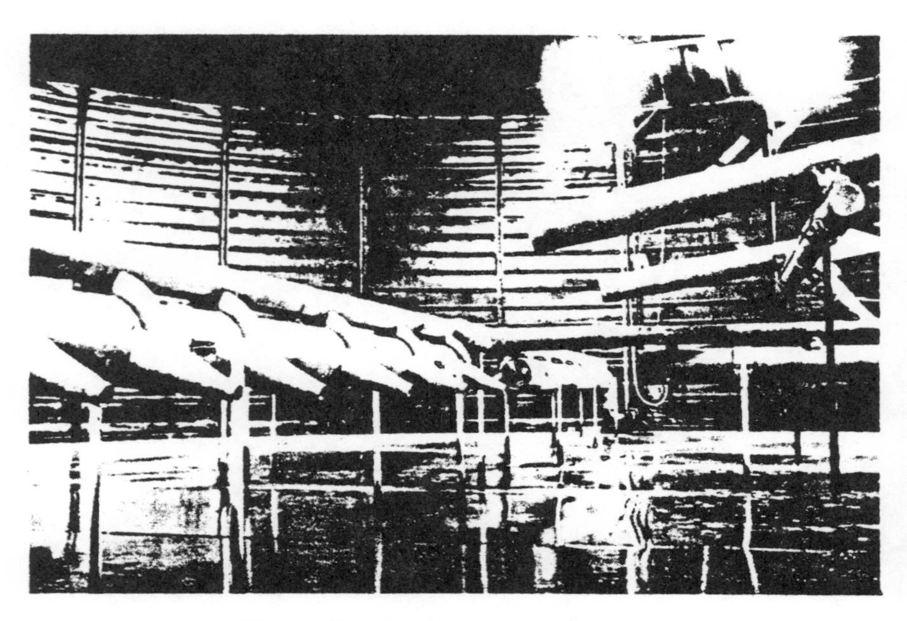

Figure 56. Sequencing batch reactor.
Photographs courtesy of CECOS International, Inc.

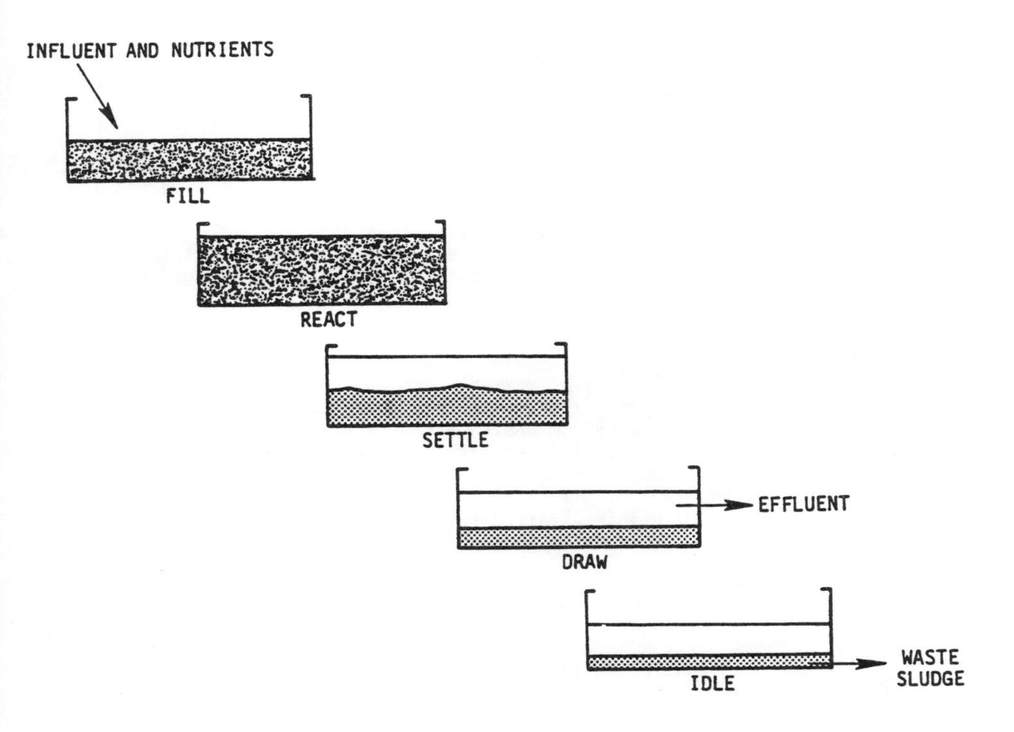

Figure 57. Five phases of treatment in the operation of a sequencing batch reactor.

Landfill in Niagara Falls, New York, and at CECOS International's commercial hazardous waste treatment, storage, and disposal facility, also in Niagara Falls (Ying et al. 1986; Staszak et al., undated).

As with any biological treatment system, the SBR can be upset by the rapid introduction of biotoxic substances. Therefore, chemical equalization of the reactor feed should be provided (Staszak et al., undated). Because of the turbid nature of the SBR effluent, filtration normally will be required as post-treatment. Final effluent polishing to meet discharge limitations can readily be accomplished by carbon adsorption (Irvine, Sojka, and Colaruotolo 1984). Management alternatives for waste activated sludge and air emissions of VOC's stripped by aeration of the leachate during the fill and/or react phases are addressed in Section 7.

Design Considerations--

The operating and cycle schedules of the SBR can be adjusted to meet specific treatment objectives at variable influent flow rates and organic loadings. In a typical 24-h treatment cycle, the fill phase might last 6 h; the react phase, 8 h; the settle phase, 5 h; the draw phase, 4 h; and the idle phase, 1 h (Staszak et al., undated). Process control is achieved through the use of automatic valves, sensors, flowmeters, timers, and microprocessors. Because the system is sensitive to rapid temperature drops, tank insulation and a supplemental heat source are required for winter operation.

Performance--

In 1983, CECOS International began laboratory-scale testing of the SBR for treatment of combined leachate/contaminated ground water/industrial wastewater. Four 10-liter SBR's were operated at different HRT's for 8 wk. Results of these early studies demonstrated that TOC could be reduced by 70 to 80 percent and that phenol could be reduced by 98 percent if the HRT was maintained between 2.5 and 10 d. In simulated cold-weather conditions, the SBR achieved discharge limitations for TOC and phenol (1000 and 2.4 mg/liter, respectively) down to 5°C if the rate of temperature decline was controlled (Herzbrun, Irvine, and Malinowski 1985; Staszak et al., undated).

CECOS's full-scale SBR demonstration facility went on-line in June 1984. During the 6-mon test period following startup, the reactor performed successfully (i.e., achieved the aforementioned discharge limitations); however, it was sensitive to rapid changes in influent quality (Staszak et al., undated).

Preliminary studies of the treatability of the Hyde Park Landfill leachate were conducted at the University of Notre Dame. Total organic carbon was reduced by about 90 percent in the small (2-liter) SBR's, which had 10-d HRT's (Irvine, Sojka, and Colaruotolo 1984). A 15-mon treatability study to determine long-term SBR performance was conducted subsequently, beginning in December 1982. Good treatment performance was consistently achieved under varying conditions of influent TOC, feed rate, aeration/mixing, HRT, MLSS concentration, organic loading, temperature, and cycle time. Experimental results from the studies involving the pilot-scale (500-liter) units are presented in Table 7. Treatment efficiency was unchanged when feeding was suspended on weekends and holidays (Ying et al. 1986).

Costs--

Figure 58 presents capital and annual O&M costs of a sequencing batch reactor, and Figure 59 presents a breakdown of these costs. The capital costs of a SBR for 25- to 100-gal/min streams range from $109,000 to $224,000; the reactor (tank) makes up about 42 percent of the capital outlay. The annual O&M costs range from $10,000 to $26,000; residuals management makes up about 44 percent of these costs. The system design is based on a 24-h treatment cycle with 10 h for filling, 10 h for reaction, 2 h for settling, and 2 h for emptying and idle time. The reactor tank design includes 100 percent redundancy in capacity to handle extra activated-sludge storage volume (Peters and Timmerhaus 1980; Guthrie 1969; Barrett 1981; U.S. Army Corps of Engineers 1984; EPA 1982a).

A SBR has obvious capital cost advantages over a conventional activated-sludge system because it eliminates the need for a separate secondary clarifier and return sludge pump. Annual O&M costs are also lower because the system is fully automated. The use of a SBR upstream of carbon adsorption can reduce the carbon requirement (and carbon regeneration costs) by 90 percent (Ying et al. 1986).

6.3.3 Powdered Activated Carbon Treatment (PACT)

Process Description--

The patented powdered activated carbon treatment (PACT^R) process (Zimpro, Inc.) involves the controlled addition of powdered activated carbon to the aeration tank of a conventional activated-sludge system. (See Subsection 6.3.1 for a description of an activated-sludge system.) Removal of organics is achieved through a combination of biological oxidation/assimilation and physical adsorption. A flow diagram of the PACT process is presented in Figure 60. Leachate is mixed with powdered activated carbon, nutrients, and biological solids. The mixed liquor is aerated for several hours to effect biological oxidation and is then discharged to a clarifier, where the powdered carbon and biological solids are settled and separated from the treated waste stream. Clarifier overflow is discharged from the PACT process for additional treatment. Clarifier underflow solids are continuously returned to the aeration tank, along with makeup carbon, to maintain the desired concentration of powdered carbon and microorganisms in the mixed liquor. Excess settled solids (consisting of powdered activated carbon, activated sludge, adsorbed organic material, and inert material) are wasted directly from the recycle stream. These waste solids are generally dewatered prior to disposal or regeneration of the activated carbon.

Applicability to Hazardous Waste Leachate--

The PACT process is applicable to nearly all wastewaters with a COD between 50 and 50,000 mg/liter (Meidl and Wilhelmi 1986). It is particularly effective for treatment of wastes such as leachate that are variable in composition and concentration, that are highly colored, and that contain refractive materials. Priority pollutants that are amenable to treatment by the PACT process are listed in Table 8. Stated advantages of the PACT process over the conventional activated-sludge system include 1) higher BOD and COD removals, 2) stability of operation with variability in influent concentration and composition, 3) enhanced removal of refractive substances and

TABLE 7. TREATABILITY OF HYDE PARK LANDFILL LEACHATE IN A 500-LITER SEQUENCING BATCH REACTOR[a] (mg/liter)

Sample[b]	TOC	COD	Phenol	HET acid	Benzoic acid	o-CBA	m-CBA	p-CBA
Influent	2000	5300	530	260	730	350	110	110
A effluent	140(93)[c]	510(90)	6(98.9)	170(35)	6(99.2)	12(97)	25(77)	3(97)
B effluent	120(94)	400(92)	1(99.8)	150(42)	2(99.7)	2(99)	3(97)	2(98)
C effluent	536(73)	1700(68)	12(97.7)	175(33)	6(99.2)	20(94)	25(77)	3(97)

[a] Source: Ying et al. 1986.

[b] Reactor A: HRT = 5 d; MLSS = 5,000 mg/liter. Reactor B: HRT = 5 d; MLSS = 10,000 mg/liter.
Reactor C: HRT = 2 d; MLSS = 10,000 mg/liter.

[c] Numbers in parentheses indicate percent reduction.

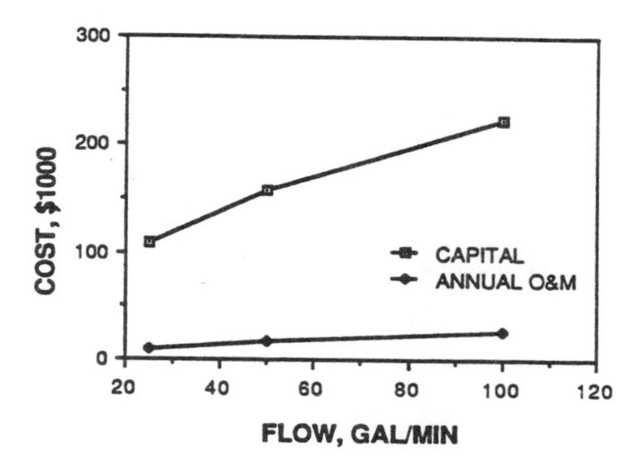

Figure 58. Capital and annual O&M costs of a sequencing batch reactor.

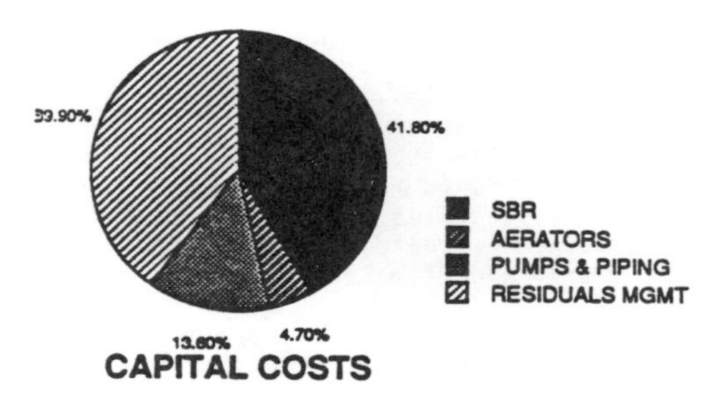

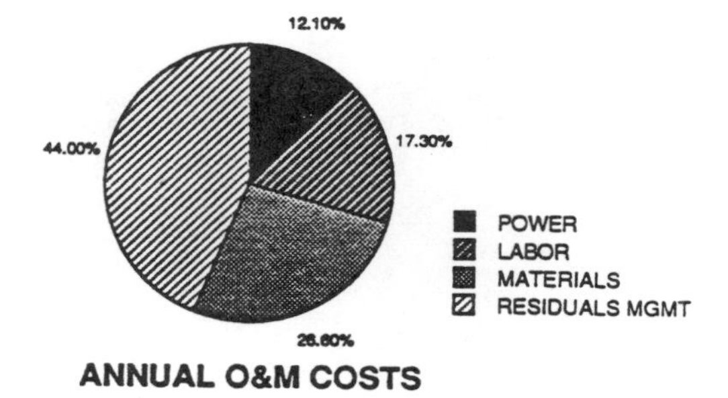

Figure 59. Breakdown of capital and annual O&M costs of a sequencing batch reactor.

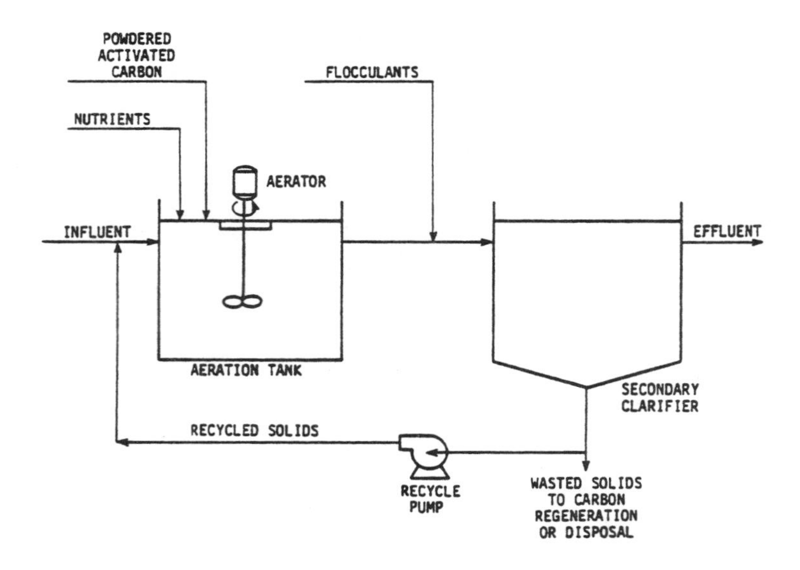

Figure 60. Schematic flow diagram of the PACT process.

priority pollutants, 4) effective color removal, 5) improved solids settling, and 6) suppression of volatilization of organics (Copa et al. 1985). Bofors-Nobel, Inc., a specialty chemicals manufacturer, recently installed a 4.5 x 10^6-gal/day PACT treatment system at a former landfill site in Muskegon, Michigan, to manage combined leachate, contaminated ground water, and industrial process wastes (Meidl and Wilhelmi 1986).

Because the PACT process is buffered against variations in organic loadings, pretreatment requirements are generally limited to neutralization; equalization has not been shown to improve performance significantly (Heath 1986). Granular-media filtration of the clarified effluent for removal of residual suspended solids may be required to meet discharge limitations. Waste solids can be processed by simple dewatering and disposal or by wet-air oxidation (described in Subsection 6.2.10) or multiple-hearth incineration for destruction of organics and regeneration of the activated carbon. For small installations, carbon regeneration is typically handled offsite. Because the PACT process limits volatilization of organics from the aeration tank, air emission controls generally are not required (Meidl and Wilhelmi 1986).

TABLE 8. PRIORITY POLLUTANTS AMENABLE TO TREATMENT BY THE PACT PROCESS[a]

Class	Compound
Volatile organics	Benzene
	Carbon tetrachloride
	Chlorobenzene
	Chloroethane
	Chloroform
	Ethylbenzene
	Methyl chloride
	Tetrachloroethylene
	Toluene
	1,1,1-Trichloroethane
	Trichloroethylene
	Trichlorofluoromethane
Acid-extractable organics	2-Chlorophenol
	2,4-Dinitrophenol
	4-Nitrophenol
	Phenol
Base/neutral-extractable organics	1,2-Dichlorobenzene
	2,4-Dinitrotoluene
	2,6-Dinitrotoluene
	Nitrobenzene
	1,2,4-Trichlorobenzene

[a] Source: Heath 1986.

Design Considerations--

The treatability of a particular leachate stream by the PACT process can be approximated from a biophysical adsorption isotherm, which expresses the quantity of material that can be assimilated/adsorbed per unit weight of carbon as a function of the effluent strength (Copa et al. 1985). Depending on the characteristics of the leachate and the treatment objectives, the mixed-liquor carbon concentration will range between 1,000 and 10,000 mg/liter (Meidl and Wilhelmi 1986). Solids residence times are typically measured in days or weeks as opposed to hours for conventional activated-sludge systems. Mixed liquor from a municipal wastewater treatment plant can be used to establish the PACT mixed liquor; however, several preliminary runs with increasing concentrations of leachate will be required to acclimate the biomass (Copa et al. 1985). Skid-mounted package treatment systems are available for low-flow applications (20,000 to 55,000 gal/day).

Performance--

Laboratory studies have shown that the PACT process is capable of better organic removal efficiencies than either activated sludge or carbon adsorption alone. This results primarily from the high mixed-liquor carbon concentrations and the long solids residence times (Heath 1986; Meidl and Wilhelmi 1986).

In 1984, Zimpro, Inc. (under contract to the California Department of Health Services) used the PACT process to conduct a pilot-scale demonstration of the treatability of leachate from the Stringfellow hazardous waste site in Glen Avon, California. Performance of the 2.4-gal/min unit, which operated at a HRT of 16 h, a SRT of 15 d, and a mixed-liquor carbon concentration of 9700 mg/liter, is summarized in Table 9. Analysis of the effluent showed it to be essentially free of priority pollutants, including benzene, chlorobenzene, dichloromethane, chloroform, 1,2-dichloroethylene, trichloroethylene, tetrachloroethylene, ethyl benzene, and toluene, that were present in the influent. Monitoring of the off-gas from the aeration tank showed the total hydrocarbon (THC) concentration to range between 2.5 and 11 ppm; these levels are comparable to background levels in that area (Copa et al. 1985).

TABLE 9. TREATABILITY OF STRINGFELLOW LEACHATE BY THE PACT PROCESS[a]

Parameter	Influent, mg/liter	Effluent, mg/liter	Removal, %
Chemical oxygen demand	1788	467	74
Biological oxygen demand	50.3	5.5	89
Dissolved organic carbon	535	154	71

[a] Source: Copa et al. 1985.

Costs--

Figure 61 presents capital and annual O&M costs of the PACT process, and Figure 62 presents a breakdown of these costs. The capital costs of PACT systems for 25- to 100-gal/min streams range from $249,000 to $492,000; the aeration basin makes up about 31 percent of the capital outlay, and the clarifier, 22 percent. The annual O&M costs range from $39,000 to $138,000; makeup carbon accounts for about 46 percent of these expenses, and residuals management, 32 percent. The design criteria for the system include a HRT of 12 h, a SRT of 10 d, and a mixed-liquor carbon concentration of 7000 mg/liter. The carbon dosage for the system is 350 mg/liter. Spent carbon is assumed to be disposed of offsite (EPA 1982a; Peters and Timmerhaus 1980; Guthrie 1969; U.S. Army Corps of Engineers 1985; Richardson Engineering Services 1979; Cran 1981).

Although the capital costs of the PACT process are higher than those of a comparable granular activated carbon adsorption system, lower annual O&M costs could favor the former process over the long term. Reportedly, the use of a single-stage PACT treatment system will result in measurable cost savings over a two-stage biological treatment/carbon adsorption system to achieve the same level of treatment performance (Meidl and Wilhelmi 1986).

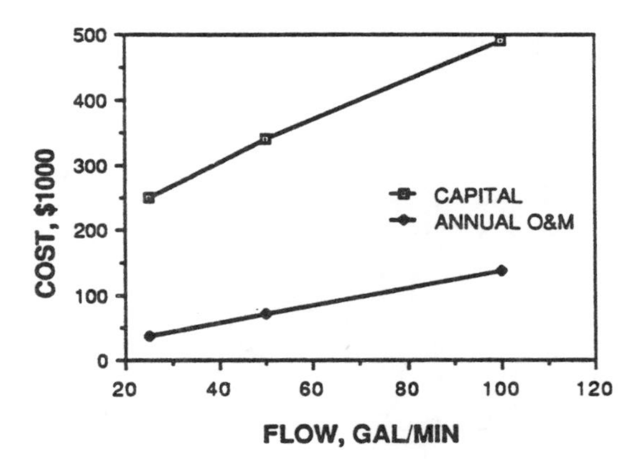

Figure 61. Capital and annual O&M costs of the PACT process.

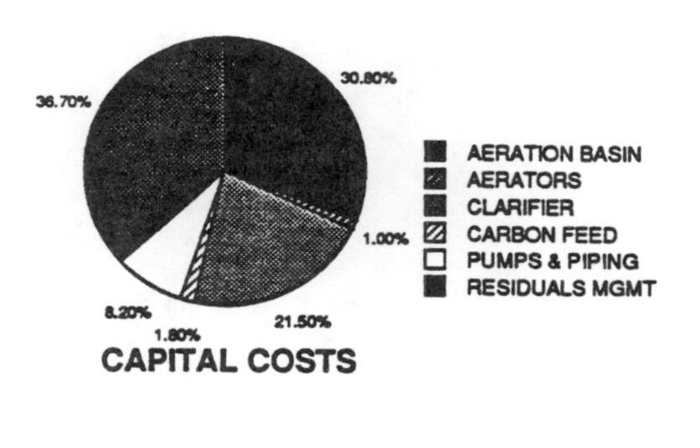

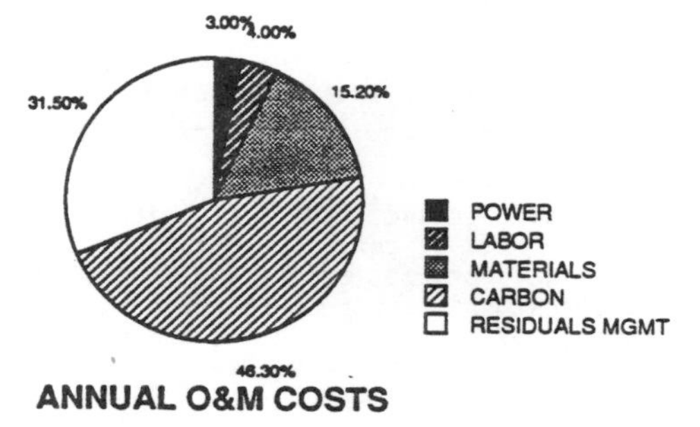

Figure 62. Breakdown of capital and annual O&M costs of the PACT process.

6.3.4 Rotating Biological Contactor

Process Description--

The rotating biological contactor (RBC) is an attached-growth, aerobic biological treatment process. An RBC consists of a series of closely spaced plastic (polystyrene, polyvinyl chloride, or polyethylene) disks on a horizontal shaft. The assemblage is mounted in a contoured-bottom tank containing leachate so that the disks are partially (about 40 percent) immersed (Figure 63). The disks, which eventually develop a slime layer 2 to 4 mm thick over the entire wetted surface, rotate slowly through the leachate and alternately contact the biomass with the organic matter in the leachate and then with the atmosphere for adsorption of oxygen. Excess biomass on the media is stripped off by rotational shear forces, and the stripped solids are held in suspension with the leachate by the mixing action of the disks. The sloughed solids are carried with the effluent to a clarifier, where they are settled and separated from the treated waste.

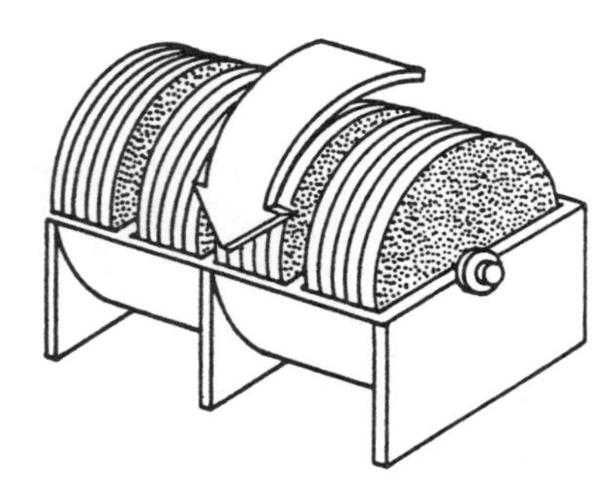

Figure 63. Schematic diagram of a rotating biological contactor.

Applicability to Hazardous Waste Leachate--

Rotating biological contactors can be used for treatment of leachate containing readily biodegradable organics. Although not as efficient as conventional activated-sludge systems, RBC's are better able to withstand fluctuating organic loadings because of the large amount of biomass they support (EPA 1982a). The proposed treatment scheme for leachate from the New Lyme Landfill in Ashtabula County, Ohio, incorporates an RBC for reduction of organic contaminants.

Like other biological processes, RBC's are either inhibited or ineffective at high concentrations of metals, refractory organics, or other toxic conditions. Equalization, metals precipitation, and neutralization should be considered minimum pretreatment requirements. Post-treatment will involve clarification for removal of biological solids and carbon adsorption for removal of residual organics. Management alternatives for sludge and air emissions of VOC's stripped from the leachate by the mixing action of the disks are addressed in Section 7.

Design Considerations--

Rotating biological contactors provide a greater degree of flexibility for meeting the changing needs of a leachate treatment plant than do other attached-growth biological processes, e.g., the trickling filter. The characteristic modular construction of RBC's permits their multiple staging to meet increases or decreases in treatment demands. The hydraulic retention time of the waste and the rotational speed of the disks can be controlled to effect the desired degree of system performance (EPA 1982a). Primary effluent from a municipal wastewater treatment plant can be used to develop a biomass on the disks; however, several preliminary runs with increasing proportions of leachate will be required to acclimate the microorganisms (Opatken, Howard, and Bond 1986). For winter operation, the units should be housed or covered, and a supplemental heat source should be provided (EPA 1982a). Packaged treatment systems that can be operated in a batch mode are available for low-flow applications.

Performance--

Pilot-scale treatability studies with leachate from the Stringfellow hazardous waste site in Glen Avon, California, were conducted at the U.S. EPA's Testing and Evaluation Facility in Cincinnati, Ohio. After a series of acclimation runs with combined leachate and primary effluent from a municipal wastewater treatment plant, the RBC was tested with lime-pretreated leachate from the site. Figure 64 shows the reduction of soluble organics with time. Effluent concentrations for DOC, BOD, and COD of 110 mg/liter (63 percent removal), 0 mg/liter (100 percent removal), and 370 mg/liter (54 percent removal), respectively, were achieved. Although biodegradable organics completely disappeared within 4 days, removal of the refractory organics from the treated effluent by adsorption onto powdered activated carbon was required to meet the discharge limits for DOC and COD (Opatken, Howard, and Bond 1986).

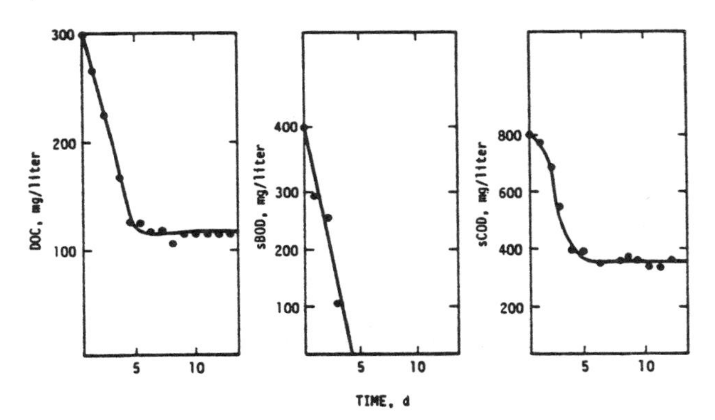

Figure 64. Treatability of Stringfellow leachate in a pilot-scale rotating biological contactor.

Source: Optaken, Howard, and Bond 1986.

Costs--

Figure 65 presents capital and annual O&M costs of a rotating biological contactor, and Figure 66 presents a breakdown of these costs. The capital costs of a RBC for 25- to 100-gal/min streams range from $103,000 to $383,000; the RBC (disks and tank) accounts for about 70 percent of the capital outlay. The annual O&M costs range from $13,000 to $36,000; residuals management accounts for about 15 percent of these expenses. Costs are based on a HRT of 150 min (EPA 1980; Richardson Engineering Services 1979; R. S. Means Co. 1986; Barrett 1981).

6.3.5 Trickling Filter

Process Description--

The trickling filter is an attached-growth, aerobic biological treatment process in which leachate is continuously distributed over a bed of rocks or plastic medium that supports the growth of microorganisms (Figure 67). The leachate trickles through the filter bed, contacts the slime layer formed on the medium, and is collected by an underdrain system. The microorganisms assimilate and oxidize substances in the leachate; as the microorganisms grow, the slime layer increases. Periodic sloughing of the slime layer into the underdrain system results from organic and hydraulic loadings on the filter, and a new slime layer begins to grow. Sloughed solids are separated from the treated effluent by settling.

Trickling filters operate under short hydraulic retention times that do not allow for complete biodegradation of organics; as a result, effluent recirculation is required to increase the net contact time of the leachate with the biomass and achieve a high organic removal efficiency. Recirculation also provides a constant hydraulic loading and dilutes high-strength leachates (EPA 1982a). Effluent recirculation is essential for trickling filters constructed with plastic medium, which has a high percentage of void space, to ensure that the medium is thoroughly wetted and will sustain microbial growth and promote effective sloughing.

Applicability to Hazardous Waste Leachate--

Trickling filters may be used to biodegrade nonhalogenated and certain halogenated organics in leachate. Although not as efficient as suspended-growth biological treatment processes, trickling filters are more resilient to variations in hydraulic and organic loadings. For this reason, trickling filters are best suited to use as "roughing" or pretreatment units that precede more sensitive processes such as activated sludge (Metcalf & Eddy 1985). The applicability of trickling filters to the full-scale treatment of hazardous waste leachate has not yet been demonstrated.

As in the case of other biological operations, pretreatment in the form of metals removal and pH adjustment should be provided. Management alternatives for sludge are addressed in Subsection 7.1. Stripping of VOC's from the leachate as it is dispersed over the filter medium should be a minor phenomenon (Metcalf & Eddy 1985).

Design Considerations--

The primary design parameters for a trickling filter include the hydraulic loading, organic loading, bed depth, and recirculation ratio. Plastic-medium filters are able to handle higher hydraulic and organic loadings, are

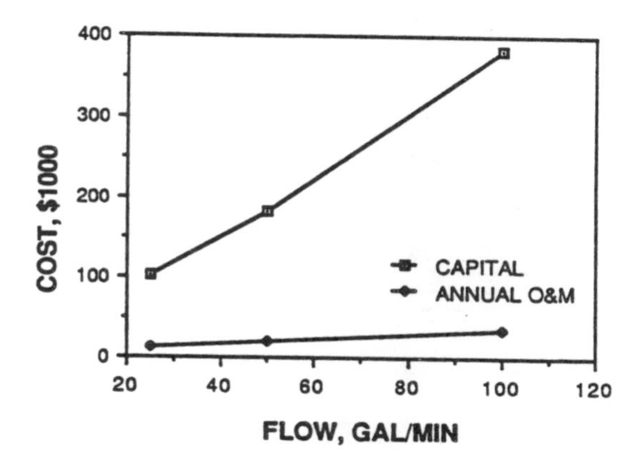

Figure 65. Capital and annual O&M costs of a rotating biological contactor.

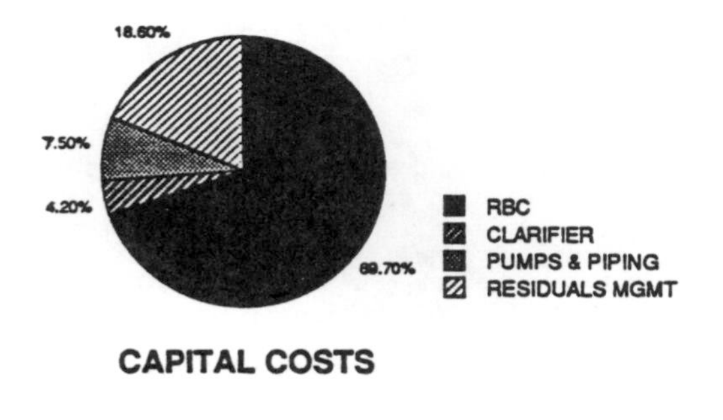

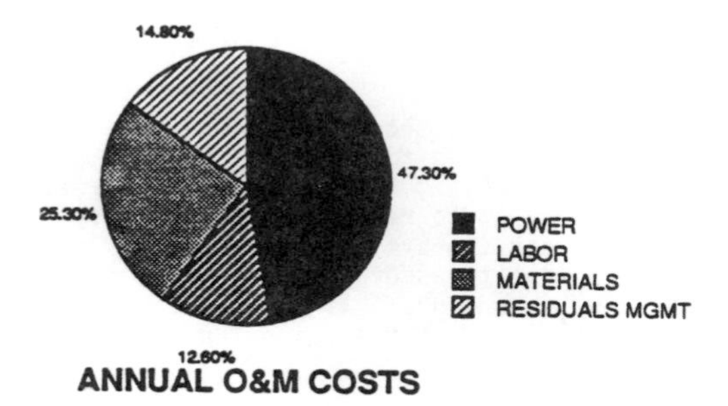

Figure 66. Breakdown of capital and annual O&M costs of a rotating biological contactor.

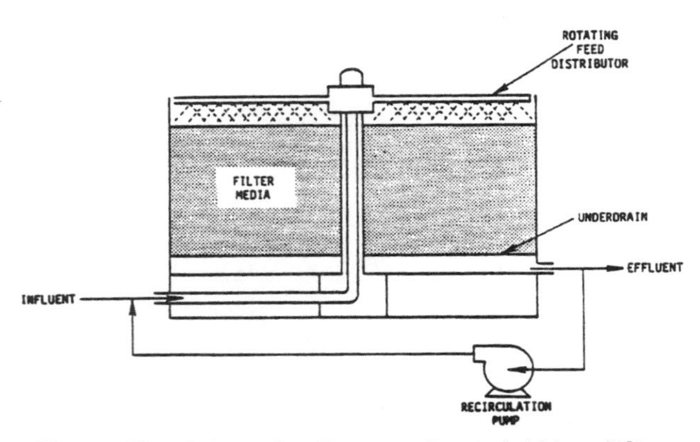

Figure 67. Schematic diagram of a trickling filter.

less susceptible to clogging, and can be built higher than rock-medium filters. Recirculation rates typically do not exceed four times the influent rate (EPA 1982a).

Performance--

When designed as roughing units, trickling filters reduce the organic load to subsequent operations and provide a more uniform feed. Process reliability is generally good; however, trickling filters are slow to recover if disrupted (EPA 1982a).

Costs--

Figure 68 presents capital and annual O&M costs of a trickling filter, and Figure 69 presents a breakdown of these costs. The capital costs of a trickling filter for 25- to 100-gal/min streams range from $150,000 to $345,000; the trickling filter (including the tower and plastic packing medium) accounts for about 27 percent of the capital outlay. The annual O&M costs range from $15,000 to $58,000; residuals management accounts for about 75 percent of these expenses. The trickling filter design is based on a hydraulic loading rate of 50 gal/day per ft^2 of filter and a recirculation ratio of 4 to 1 (Richardson Engineering Services 1979; Barrett 1981; R.S. Means Co. 1986).

6.4 POST-TREATMENT OPERATIONS

Post-treatment processes are those operations that occur downstream of the primary waste treatment stages to "polish" the system's effluent or prepare it for discharge. Such processes include filtration to remove residual suspended solids, pH adjustment to return the effluent to a neutral condition, and chlorination to disinfect the effluent prior to its discharge to surface water. Post-treatment applications of granular-media filtration and neutralization are discussed in Subsections 6.1.3 and 6.2.1, respectively. Chlorination (disinfection) is discussed in the following subsection.

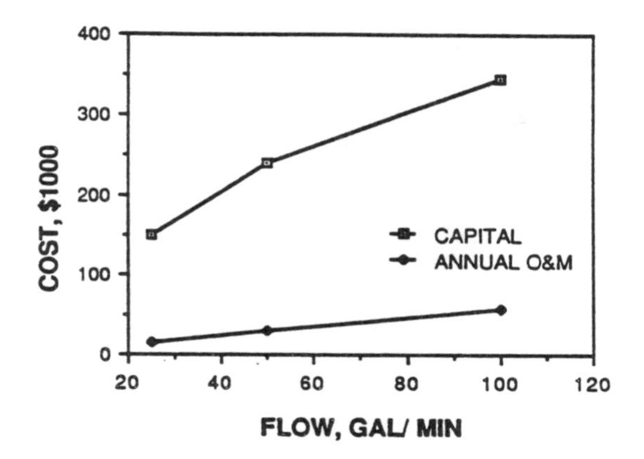

Figure 68. Capital and annual O&M costs of a trickling filter.

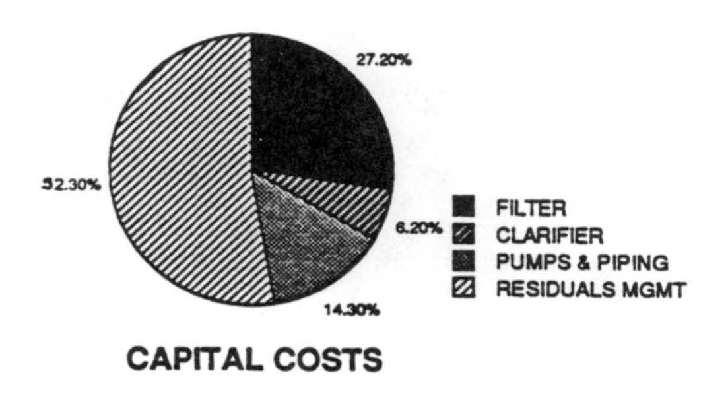

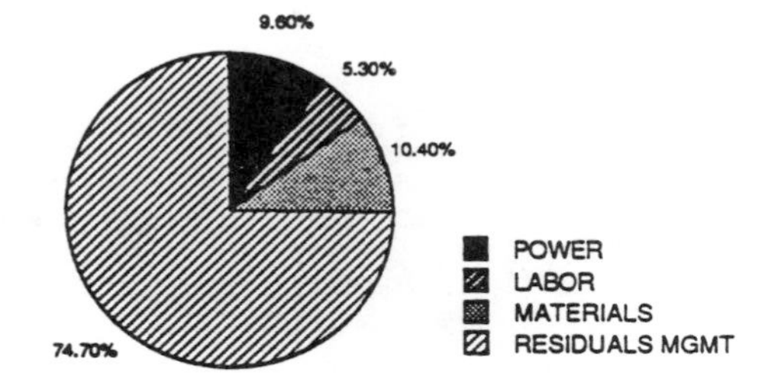

Figure 69. Breakdown of capital and annual O&M costs of a trickling filter.

6.4.1 Chlorination

Process Description--

Chlorination is a post-treatment process used primarily for disinfection to destroy microorganisms in the treated leachate prior to its discharge to ground or surface waters. The chlorination process generally involves the addition of elemental chlorine or hypochlorite salts to the final effluent. The disinfection potential of chlorine is related primarily to the strong oxidizing properties of hypochlorous acid (HOCl) and, to a lesser extent, hydrochloric acid (HCl), which are formed when chlorine combines with water. Hypochlorous acid, in turn, has a tendency to ionize to form the hypochlorite ion (OCl^-), as given below:

$$Cl_2 + H_2O \rightarrow HOCl + HCl$$

$$HOCl \rightleftarrows H^+ + OCl^-$$

The killing efficiency of HOCl is about 40 to 80 times that of OCl^- (Metcalf & Eddy 1979).

Leachate is generally contacted with chlorine in an enclosed batch tank or a continuous-flow tank. Chlorine is added to the leachate in the form of chlorine gas or calcium or sodium hypochlorite in controlled amounts through a chemical reagent feeder. As chlorine is added, it reacts with readily oxidizable substances (cyanides, organics) present in the leachate. Continued addition of chlorine past the breakpoint is required for complete disinfection.

Applicability to Hazardous Waste Leachate--

The practice of chlorination of both water supplies and wastewaters is wide-spread. The applicability of chlorination to hazardous waste leachate is limited to disinfection of the treated effluent prior to its discharge to ground or surface waters. The leachate treatment process train proposed for the Helen Kramer Landfill in Mantua Township, New Jersey, includes chlorination as the final step.

A major disadvantage of chlorination is the potential for formation of chlorinated organic compounds, some of which are known to be carcinogenic. Gaseous chlorine is toxic and requires careful handling; hypochlorite salts are much simpler and safer to use.

Design Considerations--

The effectiveness of chlorination for disinfection depends on pH, temperature, contact time, mixing, and the presence of interfering compounds (EPA 1982a). Temperature and pH affect the extent to which hypochlorous acid ionizes to form the weaker oxidizing hypochlorite ion. For maximum disinfection potential of the system, the pH should be maintained below 7.5. Contact time and mixing affect the degree of exposure of microorganisms to the disinfectant. Generally, a retention time of 15 to 30 minutes is required. Rapid initial mixing and the use of a baffled contact tank can prevent short-circuiting (Figure 70). Baffling can be either the over-and-under or the end-around variety. The presence of readily oxidizable constituents in the treated leachate increases the chlorine demand and the required chemical dosage.

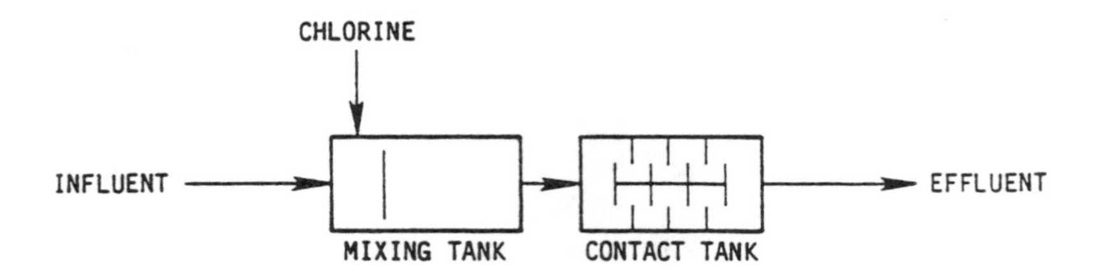

Figure 70. Schematic flow diagram of a chlorination system.

The size of the contact tank is based primarily on the chlorine demand of the water and on its flow rate. Automated feed control systems are used to adjust the chlorine dosage to changes in the influent characteristics. The more sophisticated of these control systems use a residual chlorine analyzer, which controls the chlorine dosage based on residual chlorine levels in the effluent.

Performance--

Chlorination is an effective means of disinfection and is less costly than alternatives such as ozonation. Performance of chlorination systems is tied to the reliability of automated control systems.

Costs--

Figure 71 presents capital and annual O&M costs of a chlorination system, and Figure 72 presents a breakdown of these costs. The capital costs of chlorination for 25- to 100-gal/min streams range from $31,000 to $71,000; the mixing tank makes up about 20 percent of the capital outlay; the contact tank, 30 percent; and the chlorine feed system, 50 percent. The annual O&M costs range from $5,000 to $7,000; chemicals make up about 19 percent of these expenses. Estimates of chemical costs are based on the use of chlorine gas (Richardson Engineering Services 1979).

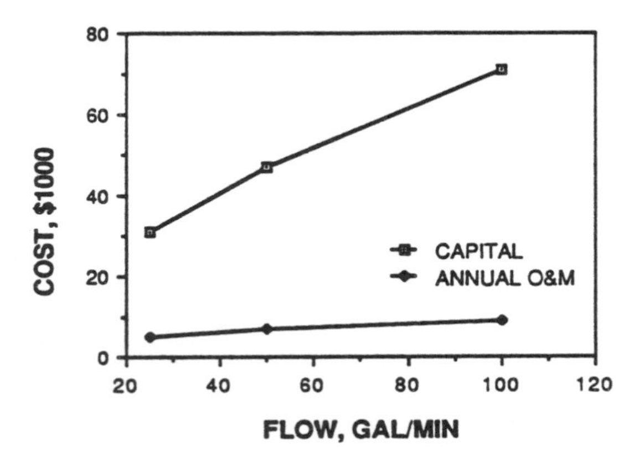

Figure 71. Capital and annual O&M costs of a chlorination system.

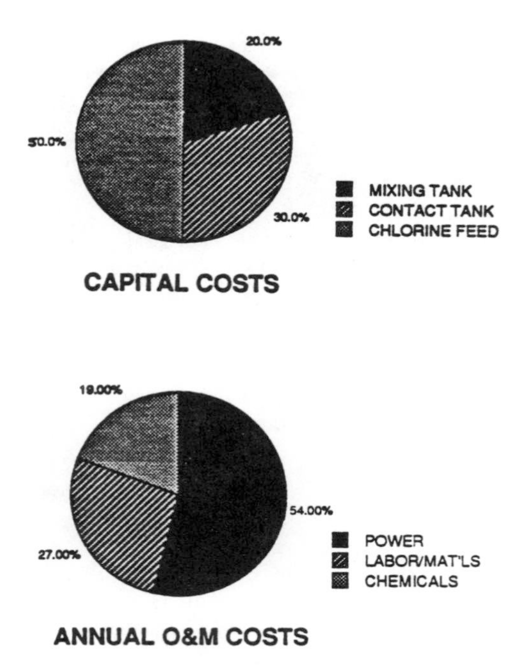

Figure 72. Breakdown of capital and annual O&M costs of a chlorination system.

7. Residuals Management

Important considerations in the selection of a leachate treatment process are the type and volume of residuals generated by the process, as these factors affect operating and maintenance costs. Residuals generated by the treatment processes described in Section 6 include sludge, air emissions, concentrated liquid waste streams, and spent carbon. Table 10 presents a listing of the residuals generated by each of these processes. Current residuals management practices are discussed under the appropriate headings in the remainder of this section.

TABLE 10. RESIDUALS GENERATED BY THE VARIOUS LEACHATE TREATMENT PROCESSES

	Residuals			
Treatment process	Sludge	Air emissions	Concentrated liquid waste stream	Spent carbon
Pretreatment operations				
Sedimentation	x			
Granular-media filtration			x	
Oil/water separation			x	
Physical/chemical treatment operations				
Neutralization	x			
Precipitation/flocculation/sedimentation	x			
Oxidation/reduction	x			
Carbon adsorption			x	x
Air stripping		x		
Steam stripping			x	
Reverse osmosis			x	
Ultrafiltration			x	
Ion exchange			x	
Wet-air oxidation		x		
Biological treatment operations				
Activated sludge	x	x		
Sequencing batch reactor	x	x		
Powdered activated carbon treatment (PACT)	x			x
Rotating biological contactor	x	x		
Trickling filter	x			

7.1 SLUDGE

Physical/chemical treatment sludges are generated by the sedimentation of suspended solids and/or insoluble reaction byproducts. Biological treatment sludges are generated by the microbial conversion of soluble organics to cellular biomass. Because contaminants are often concentrated in these sludges, they will require further treatment and disposal in an environmentally sound manner.

7.1.1 Sludge Dewatering

Sludge dewatering is a physical (mechanical) operation used to reduce the moisture content and volume of sludges. Moisture reduction, which is normally required prior to the landfilling or incineration of sludges, facilitates handling and reduces transportation and ultimate disposal costs.

Popular dewatering methods include filter-press dewatering, vacuum filtration, belt-press dewatering, centrifugation, and evaporation. Filter-press dewatering is the most common method used to manage sludges generated by leachate treatment operations. This method is currently practiced at the Stringfellow hazardous waste site. Filter presses consist of a series of vertical recessed plates held rigidly in a frame and pressed together between fixed and moving ends. The sludge fed into the press flows into the cavities between adjacent plates. Water is forced through the filter cloths covering the face of each plate at 100 to 200 lb/in^2, and the solids are retained on the surface of the cloth. The filtrate is collected in drainage ports at the bottom of each press chamber and returned to the head of the treatment process. The press is then opened and the individual plates are removed to allow the filter cake to fall into a sludge container.

Filtration is suitable for dewatering both biological and chemical sludges. The effectiveness of dewatering for a particular application depends on the type of filter, particle size distribution, and the solids content of the sludge. Filter-press dewatering is somewhat limited by the relatively short life of the filter cloths and the fact that presses must be installed well above the floor level to allow sufficient room for the filter cake to drop out onto conveyors or into containers.

7.1.2 Chemical Stabilization/Solidification

Chemical stabilization/solidification involves the addition of absorbents and solidifying agents to the sludge. This process is designed to improve the handling and physical characteristics of the sludge, to decrease the surface area for transport of hazardous constituents, to limit the solubility of pollutants in the sludge, and/or to detoxify the contained pollutants (EPA 1982b).

Several methods are currently available for the stabilization/solidification of sludges. These include sorption, lime/fly-ash/pozzolan processes, pozzolan/portland cement processes, thermoplastic microencapsulation, and macroencapsulation (jacketing) (EPA 1986).

Selection of a stabilization/solidification method depends on many factors: characteristics of the sludge, applicability of a particular method, cost, sludge volume, degree of hazard, required pretreatment, disposal-site characteristics, handling characteristics of the stabilized/solidified sludge, and the specifications of the sludge to be disposed of. The advantages and disadvantages of each method must be evaluated carefully before a particular method is selected.

Pretreatment of some sludges is advisable before stabilization/solidification. Pretreatment removes or alters interfering compounds, such as acids and oxidizers, that react with solidification reagents. The use of processes such as settling and dewatering can also reduce the volume of sludge to be solidified. The addition of chemical binders can remove toxic compounds from solution and retain them in solids. Various techniques, such as bulking and homogenizing sludges, can improve the efficiency of the stabilization/solidification process.

Stabilization/solidification methods are most amenable to inorganic materials in aqueous solution or suspension that contain toxic metals or inorganic salts. Organic compounds interfere with the physical and chemical processes that are instrumental in binding the waste together; therefore, organic wastes are not amenable to this treatment technology (EPA 1982b).

7.1.3 Biological Stabilization

Stabilization of biological sludges can be achieved by aerobic or anaerobic sludge digestion. In aerobic digestion, microorganisms in the presence of oxygen consume to depletion the available food in the sludge and then continue to feed on their own protoplasm to continue living. Most of the cell tissue is oxidized to carbon dioxide, water, and ammonia; the remaining tissue is composed of inert components and nonbiodegradable organic compounds (Metcalf & Eddy 1979). The product of aerobic digestion is an odorless, humuslike, biologically stable sludge that is very amenable to dewatering. In anaerobic sludge digestion, organic material is biologically converted under anaerobic conditions to methane and carbon dioxide. The digestion process takes place in an airtight reactor over varying periods of time and produces a nonputrescible stabilized sludge.

7.1.4 Incineration

Incineration is another alternative for the disposition of chemical or biological sludges. Depending on the characteristics of the sludge, dewatering may be the only pretreatment required. In the interest of minimizing transportation and disposal costs, sludge should be dewatered to approximately 30 percent solids or whatever solids content is required by the incineration facility. Less water content in the sludge will also reduce incineration costs.

Incineration of sludge is an alternative to stabilization and solidification when organic wastes are involved, as these wastes are not amenable to stabilization/solidification. Another advantage of incineration is that it destroys some or all of the hazardous constituents or characteristics of the sludge. The disadvantages of incineration include high operating and maintenance costs compared with those of other treatment/disposal options and the generation of residues (such as ash) and air emissions, which must be managed in accordance with environmental regulations.

7.1.5 Land Disposal

For sludge to be accepted for disposal at a hazardous waste landfill, the solids content must meet or better State standards, and it must not contain any free liquids. If the sludge demonstrates hazardous characteristics or is a hazardous waste by definition, it must be disposed of at an EPA-approved hazardous waste landfill.

Some sludges proposed for RCRA land disposal ban as early as 1988 must be treated in accordance with EPA treatment standards prior to their disposal. These required treatment practices will have an impact on the cost of disposal. Facilities affected by the land disposal ban must consider the costs associated with the various sludge treatment/disposal options before selecting a method.

7.2 AIR EMISSIONS

By design, certain leachate treatment technologies (e.g., air stripping) transfer VOC's from the liquid phase to the vapor phase. Other treatment processes (e.g., activated sludge, rotating biological contactor, and sequencing batch reactor) strip some VOC's by nature of the aeration process. Unless provisions are made for treatment of air emissions, VOC's will be discharged to the atmosphere.

7.2.1 Vapor-Phase Carbon Adsorption

Carbon adsorption is a very effective method for removing VOC's from the vapor phase. Contaminant-laden air is passed through a column of activated carbon. Organics are adsorbed from the air stream, and clean air is discharged to the atmosphere. After a period of 3 to 6 months (depending on the level of contamination and the volume of carbon), the carbon column becomes exhausted and must be exchanged for new carbon. The spent carbon is returned to the supplier for regeneration or disposed of offsite. Four of the six case-study sites presented in Table 4 (p. 21) that remove volatile organics from leachate by air stripping control VOC emissions by vapor-phase carbon adsorption.

7.2.2 Fume Incineration

Fume incineration may be useful for the control of combustible atmospheric emissions that are generated by air stripping of organic compounds from leachate. A fume incinerator is an enclosed refractory-lined chamber with a fuel burner in one end capable of heating the chamber to about 1500°F. The chamber is sized to provide a retention time of about 0.5 second. In practice, the burner is fired with fuel oil or gas and the chamber is preheated to the desired operating temperature. The contaminated air, which is then injected at the burner end, provides some of the combustion air needed by the fuel.

The incinerator operates with about 50 percent excess air. The high temperature and excess oxygen in the air destroy the organic compounds that are emitted with the fuel combustion products, largely as CO_2 and H_2O. Some carbon monoxide, unburned organics, nitrogen oxides, and (if any chlorides are present) hydrogen chloride are also emitted along with nitrogen and excess oxygen.

Fume incineration is a well-established technology, and properly designed and operated fume incinerators can reduce organic emissions by at least 80 percent. Hydrogen chloride emissions present a potential problem if large quantities of chlorinated organic compounds are present in the contaminated air, and they may require control with a scrubber. Fuel is the major operating expense, as the contaminated air is cool and the organic contaminants are at too low a level to have any significant fuel value. Heat recovery could be considered to preheat the incoming air (and thereby reduce fuel costs) or to generate steam.

The Sylvester Site in Nashua, New Hampshire, uses an oil-fired boiler as a fume incinerator to treat air emissions from a ground-water/leachate air-stripping process. This air stream contains primarily tetrahydrofuran, methyl ethyl ketone, butyl alcohol, toluene, and smaller amounts of other organics. The desired destruction efficiency is 99.99 percent.

7.3 CONCENTRATED LIQUID WASTE STREAMS

Liquid waste streams (backwash water, concentrate, and condensate) generated by many physical/chemical treatment operations contain high concentrations of suspended solids or pollutants that the particular treatment process was designed to remove. Backwash water is usually returned to the head works of the treatment plant; however, if recycling is not practiced, the backwash water must be treated or disposed of. Options available for the treatment or disposal of concentrates and condensates include incineration or stabilization/solidification in preparation for land disposal.

7.4 SPENT CARBON

Granular and powdered activated carbon are used extensively for leachate treatment and for the control of air pollutants such as VOC's. When the carbon becomes exhausted, it can either be regenerated and reused or disposed of by incineration or land disposal. In most cases, however, spent carbon is regenerated by the supplier and reused.

7.4.1 Carbon Regeneration

Carbon regeneration techniques can be categorized as either thermal regeneration or nondestructive regeneration processes. Thermal oxidation involving the use of a multiple-hearth, fluidized-bed, or rotary kiln furnace is the most prevalent means of regenerating granular activated carbon. Wet-air oxidation can be used for thermal regeneration of powdered activated carbon. One drawback to thermal regeneration is the 5 to 10 percent carbon loss attributable to a combination of handling losses and carbon destroyed by the process. Also, regenerated carbon exhibits different performance characteristics, and removal efficiencies are less than those of virgin activated carbon. When carbon is contaminated with a compound that requires extremely high temperatures for desorption, thermal regeneration may not be possible. Consequently, incineration or land disposal would have to be considered.

Nondestructive regeneration of activated carbon is accomplished by the use of steam to remove VOC's, solvents to remove a variety of organics, and a pH shift for weak acids and bases. Steam regeneration does not physically change the activated carbon and it minimizes carbon losses. If desired, steam regeneration can be used for the recovery of organics from the spent carbon. The steam regeneration process generates a small volume of concentrated aqueous waste that must be handled as described in Subsection 7.3.

7.4.2 Incineration/Land Disposal

If spent carbon cannot be regenerated because it is contaminated with a compound that requires extremely high temperatures for desorption or because regeneration is cost-prohibitive, the spent carbon must be incinerated or disposed of. If the carbon is used to remove hazardous substances, it will have to be disposed of at an approved hazardous waste disposal facility.

References

Applegate, L. E. 1984. Membrane Separation Processes. Chemical Engineering, 91(12):64-89.

Ball, W. P., M. D. Jones, and M. C. Kavanaugh. 1984. Mass Transfer of Volatile Organic Compounds in Packed Tower Aeration. Journal of the Water Pollution Control Federation, 56(2):127-136.

Barrett, O. H. 1981. Installed Cost of Corrosion-Resistant Piping. Chemical Engineering, 88(22):100-101.

Blaney, B. L. 1986. Alternative Techniques for Managing Solvent Wastes. Journal of the Air Pollution Control Association, 36(3):275-285.

Canter, L. W., and R. C. Knox. 1986. Ground Water Pollution Control. Lewis Publishers, Inc., Chelsea, Michigan.

Chian, E. S. K., and F. B. DeWalle. 1977. Evaluation of Leachate Treatment, Volume II. Biological and Physical-Chemical Processes. EPA/600/2-77/186b.

Copa, W. M., et al. 1985. Powdered Activated Carbon Treatment (PACT™) of Leachate From the Stringfellow Quarry. Presented at the 11th Annual EPA/HWERL Research Symposium, April 30, 1985, Cincinnati.

Cran, J. 1981. Improved Factored Methods Give Better Preliminary Cost Estimates. Chemical Engineering, 88(7):73.

DeWolf, G., R. Hearn, and P. Storm. 1982. Cost of Environmental Control Technologies. Granular Activated Carbon Applications in Water and Wastewater Treatment. Prepared for the U.S. Environmental Protection Agency by Radian Corporation under Contract No. 68-03-3038.

Dietrich, M. J., T. L. Randall, and P. J. Canney. 1985. Wet Air Oxidation of Hazardous Organics in Wastewater. Environmental Progress, 4(3):171-177.

Guthrie, K. M. 1969. Capital Cost Estimating. Chemical Engineering, March 24, 1969, pp. 114-142.

Hafslund, E. R. 1979. Distillation. In: Kirk-Othmer Encyclopedia of Chemical Technology. 3rd ed. Vol. 7. John Wiley & Sons, New York.

Hansen, W. G., and H. L. Rishel. Undated. Cost Comparisons of Treatment and Disposal Alternatives for Hazardous Wastes. Vol. 1. Prepared for the U.S. Environmental Protection Agency, Municipal Environmental Research Laboratory, Cincinnati.

Heath, Jr., H. W. 1986. Update on the PACT Process. In: Proceedings of the National Conference on Hazardous Wastes and Hazardous Materials, March 4-6, 1986, Atlanta.

Heister, N. K. 1973. Ion Exchange. In: Perry's Chemical Engineers' Handbook, R. H. Perry and C. H. Chilton, eds. 5th ed. McGraw-Hill Book Co., New York.

Henry, J. D., et al. 1984. Novel Separation Processes. In: Perry's Chemical Engineers' Handbook, R. H. Perry and D. Green, eds. 6th ed. McGraw-Hill Book Co., New York.

Herzbrun, P. A., R. L. Irvine, and K. C. Malinowski. 1985. Biological Treatment of Hazardous Waste in Sequencing Batch Reactors. Journal of the Water Pollution Control Federation, 51(12):1163-1167.

Irvine, R. L., S. A. Sojka, and J. F. Colaruotolo. 1984. Enhanced Biological Treatment of Leachates From Industrial Landfills. Hazardous Waste, 1(1):123-135.

Johnson, J. S. 1982. Reverse Osmosis. In: Kirk-Othmer Encyclopedia of Chemical Technology. 3rd ed. Vol. 20. John Wiley & Sons, New York.

Johnson, T., F. Lenzo, and K. Sullivan. 1985. Raising Stripper Temperature Raises MEK Removal. Pollution Engineering, 17(9):34-35.

Kavanaugh, M. C., and R. R. Trussell. 1980. Design of Aeration Towers to Strip Volatile Contaminants From Drinking Water. Journal of the American Water Works Association, 72(12):684-692.

McCabe, W. L., and J. C. Smith. 1976. Unit Operations of Chemical Engineering. 3rd ed. McGraw-Hill Book Co., New York.

McCoy & Associates. 1985. Membrane Separation Technology: Applications to Waste Reduction and Recycling. The Hazardous Waste Consultant, 3(3):4-1 to 4-27.

McCoy & Associates. 1986. A Guide to Innovative Hazardous Waste Treatment Processes. The Hazardous Waste Consultant, 4(1):4-1 to 4-21.

Meidl, J. A., and A. R. Wilhelmi. 1986. PACT™ Tackles Contaminated Groundwater and Leachate. Hazardous Materials and Waste Management, 4(5):8+.

Metcalf & Eddy, Inc. 1979. Wastewater Engineering: Treatment, Disposal, Reuse. 2nd ed. McGraw-Hill Book Co., New York.

Metcalf & Eddy, Inc. 1985. Briefing: Technologies Applicable to Hazardous Waste. Prepared for the U.S. Environmental Protection Agency, Office of Research and Development, Hazardous Waste Engineering Research Laboratory, Cincinnati.

Metry, A., and F. L. Cross, Jr. 1976. Leachate Control and Treatment. Vol. 7, Environmental Monograph Series. Technomic Publishing Co., Westport, Connecticut.

Mikucki, W. J., et al. 1981. Characteristics, Control and Treatment of Leachate at Military Installations. U.S. Army Construction Engineering Research Laboratory, CERL-IR-N-97.

Miller, S. A., et al. 1984. Liquid-Solid Systems. In: Perry's Chemical Engineers' Handbook, R. H. Perry and D. Green, eds. 6th ed. McGraw-Hill Book Co., New York.

Opatken, E. J., H. K. Howard, and J. J. Bond. 1986. Stringfellow Leachate Treatment With RBC. Presented at the AIChE National Meeting, November 4, 1986, Miami.

Patterson, J. W. 1985. Industrial Wastewater Treatment Technology. 2nd ed. Butterworth Publishers, Boston.

Peters, S. P., and K. D. Timmerhaus. 1980. Plant Design and Economics for Chemical Engineers. 3rd ed. McGraw-Hill Book Co., New York.

R. S. Means Co., Inc. 1986. Means Mechanical Cost Data. 9th ed. R. S. Means Co., Inc., Kingston, Massachusetts.

Richardson Engineering Services, Inc. 1979. Process Plant Construction Estimating Standards. Vol. 4. Process Equipment. Richardson Engineering Services, Inc., San Marcos, California.

Rothman, D. W., J. C. Gorton, Jr., and J. A. Sanford. 1984. Remedial Investigation and Feasibility Study for the Pollution Abatement Services Site (Oswego, New York). In: Proceedings of the 5th National Conference on Management of Uncontrolled Hazardous Waste Sites, November 7-9, 1984, Washington, D.C.

Seader, J. D., and Z. M. Kurtyka. 1984. Distillation. In: Perry's Chemical Engineers' Handbook, R. H. Perry and D. Green, eds. 6th ed. McGraw-Hill Book Co., New York.

Shuckrow, A. J., A. P. Pajak, and C. J. Touhill. 1982. Management of Hazardous Waste Leachate. U.S. Environmental Protection Agency, SW-871.

Slater, C. S. 1982. Concentration and Separation Technologies for the Treatment of High Strength Industrial Wastewaters and Leachate. EPA Progress Report No. 14.

Soffel, R. W. 1978. Activated Carbon. In: Kirk-Othmer Encyclopedia of Chemical Technology. 3rd ed. Vol. 4. John Wiley & Sons, New York.

Staszak, C. N., et al. Undated. Full-Scale Sequencing Batch Reactor Use in a Commercial Facility.

Syzdek, A. C., and R. C. Ahlert. 1984. Separation of Landfill Leachate With Polymeric Ultrafiltration Membranes. Journal of Hazardous Materials, 9(2):209-220.

Turner, R. G. 1986. A Review of Treatment Alternatives for Wastes Containing Nonsolvent Halogenated Organics. Journal of the Air Pollution Control Association, 36(3):728-737.

U.S. Army Corps of Engineers. 1985. Computer-Aided Cost Estimating System. Vol. 3. Unit Prices. U.S. Army Corps of Engineers, Huntsville, Alabama.

U.S. Environmental Protection Agency. 1977. Process Design Manual. Wastewater Treatment Facilities for Sewered Small Communities. EPA/625/1-77/009.

U.S. Environmental Protection Agency. 1980. Innovative and Alternative Technology Assessment Manual. EPA/430/9-78/009.

U.S. Environmental Protection Agency. 1982a. Handbook: Remedial Action at Waste Disposal Sites. EPA/625/6-82/006.

U.S. Environmental Protection Agency. 1982b. Guide to the Disposal of Chemically Stabilized and Solidified Waste (Revised). SW 872.

U.S. Environmental Protection Agency. 1983. Superfund Record of Decision (EPA Region 1). Sylvester Site, Nashua, New Hampshire. EPA/ROD/R01-83/007.

U.S. Environmental Protection Agency. 1984a. Superfund Record of Decision (EPA Region 2). Pollution Abatement Services (PAS) Site, Oswego Site, New York. EPA/ROD/R02-84/008.

U.S. Environmental Protection Agency. 1984b. Superfund Record of Decision (EPA Region 9). Stringfellow Acid Pits Site. Glen Avon, California. EPA/ROD/R09-84/007.

U.S. Environmental Protection Agency. 1984c. Superfund Record of Decision (EPA Region 3). Tyson's Dump, Upper Merion Township, Pennsylvania. EPA/ROD/R03-84/008.

U.S. Environmental Protection Agency. 1985a. Guidance on Remedial Investigations Under CERCLA. EPA/540/G-85/002.

U.S. Environmental Protection Agency. 1985b. Superfund Record of Decision (EPA Region 2). Gloucester Environmental Management Services Landfill, Gloucester Township, Camden County, New Jersey. EPA/ROD/R02-85/019.

U.S. Environmental Protection Agency. 1985c. Superfund Record of Decision (EPA Region 2). Helen Kramer Landfill, Mantua Township, New Jersey. EPA/ROD/R02-85/020.

U.S. Environmental Protection Agency. 1985d. Superfund Record of Decision (EPA Region 3). Heleva Landfill Site, North Whitehall, Pennsylvania. EPA/ROD/R03-85/011.

U.S. Environmental Protection Agency. 1985e. Superfund Record of Decision (EPA Region 2). Lipari Landfill, Mantua Township, New Jersey. EPA/ROD/R02-85/023.

U.S. Environmental Protection Agency. 1985f. Superfund Record of Decision (EPA Region 5). New Lyme, Ashtabula County, Ohio. EPA/ROD/R05-85/023.

U.S. Environmental Protection Agency. 1985g. Superfund Record of Decision (EPA Region 3). Sand, Gravel, and Stone Site, Elkton, Civil County, Maryland. EPA/ROD/R03-85/015.

U.S. Environmental Protection Agency. 1985h. Handbook: Remedial Action at Waste Disposal Sites (Revised). EPA/625/6-85/006.

U.S. Environmental Protection Agency. 1986. Stabilization/Solidification of Hazardous Waste. EPA/600/D-86/028.

Warner, H. P., J. M. Cohen, and J. C. Ireland. 1980. Determination of Henry's Law Constants of Selected Priority Pollutants. Prepared by the Wastewater Research Division, Municipal Environmental Research Laboratory, Office of Research and Development, U.S. Environmental Protection Agency, Cincinnati, Ohio.

Whittaker, H. 1984. Development of a Mobile Reverse Osmosis Unit for Spill Cleanup. In: Proceedings of the 1984 Hazardous Materials Spills Conference, April 9-12, 1984, Nashville.

Ying, W., et al. 1986. Biological Treatment of a Landfill Leachate in Sequencing Batch Reactors. Environmental Progress, 5(1):41-50.

Zimpro, Inc. 1984. Zimpro Sludge Management Systems. Manual 300. Zimpro, Inc., Rothschild, Wisconsin.